Cybernetics and the Origin of Information

GROUNDWORKS

Series Editors:

Arne De Boever, California Institute of the Arts

Bill Ross, Staffordshire University

Jon Roffe, University of New South Wales

Ashley Woodward, University of Dundee

What are the hidden sources that determine the contemporary moment in continental thought? This series goes "back to the source," publishing English translations of the hidden origins of our contemporary thought in order to better understand not only that thought, but also the world it seeks to understand. The series includes important French, German, and Italian texts that form the lesser-known background to prominent work in contemporary continental philosophy. With an eye on the contemporary moment—on both world-historical events and critical trends—Groundworks seeks to recover foundational but forgotten texts and to produce a more profound engagement not only with the contemporary but also with the sources that have shaped it.

The Dialectic of Duration, by Gaston Bachelard
    Translated and annotated by Mary McAllester Jones
    Introduction by Cristina Chimisso
The Birth of Physics, by Michel Serres
    Translated by David Webb and Bill Ross
The Genesis of Living Forms, by Raymond Ruyer
    Translated by Jon Roffe and Nicholas B. de Weydenthal
Cybernetics and the Origin of Information, by Raymond Ruyer
    Translated by Amélie Berger-Soraruff, Andrew Iliadis, Daniel W. Smith, and Ashley Woodward

Cybernetics and the Origin of Information

Raymond Ruyer

Translated by

Amélie Berger-Soraruff
Andrew Iliadis
Daniel W. Smith
Ashley Woodward

ROWMAN & LITTLEFIELD
Lanham • Boulder • New York • London

Published by Rowman & Littlefield
An imprint of The Rowman & Littlefield Publishing Group, Inc.
4501 Forbes Boulevard, Suite 200, Lanham, Maryland 20706
www.rowman.com

86-90 Paul Street, London EC2A 4NE

Copyright © 2024 by The Rowman & Littlefield Publishing Group, Inc.
Originally published as *La Cybernetique et l'origine de l'Information* in French by Flammarion, Paris, 1954. Second edition, 1967.

All rights reserved. No part of this book may be reproduced in any form or by any electronic or mechanical means, including information storage and retrieval systems, without written permission from the publisher, except by a reviewer who may quote passages in a review.

British Library Cataloguing in Publication Information Available

Library of Congress Cataloging-in-Publication Data

Names: Ruyer, Raymond, 1902–1987, author. | Berger-Soraruff, Amélie, translator.
Title: Cybernetics and the origin of information / Raymond Ruyer ; translated by Amélie Berger-Soraruff, Andrew Iliadis, Daniel W. Smith, Ashley Woodward.
Other titles: Cybernétique et l'origine de l'information. English
Description: Lanham : Rowman & Littlefield, [2024] | Series: Groundworks | Translation of: Cybernétique et l'origine de l'information. | Includes bibliographical references and index.
Identifiers: LCCN 2023038781 (print) | LCCN 2023038782 (ebook) | ISBN 9781786614971 (cloth) | ISBN 9781786614988 (paperback) | ISBN 9781786614995 (ebook)
Subjects: LCSH: Cybernetics. | Information theory.
Classification: LCC Q175 .R8913 2024 (print) | LCC Q175 (ebook) | DDC 003/.5—dc23/eng/20231016
LC record available at https://lccn.loc.gov/2023038781
LC ebook record available at https://lccn.loc.gov/2023038782

Contents

Translator's Introduction: Raymond Ruyer and the Philosophy of Information   vii
   Ashley Woodward

Note on the Translation   xxv

Introduction   1

1 The Main Types of Information Machines   19

2 Framing Activities and Framed Mechanisms   47

3 The Space of Behavior and Axiological "Space"   61

4 Communication   73

5 The Origin of Information   81

6 Negative Anti-Chance and Positive Anti-Chance   93

7 Past-Future and Cybernetics   113

8 The Mixed Origin of Information   129

9 Summary and Conclusion (To the First Edition)   141

10 The Problems of Cybernetics in 1967   143

Notes   183

Bibliography   199

Index   205

About the Author and Translators   213

Translator's Introduction
Raymond Ruyer and the Philosophy of Information

Ashley Woodward

Raymond Ruyer has a fair claim to being one of the very first philosophers of information, in the contemporary sense of the term.[1] This sense of "information," as something quantifiable, physical, and mechanical, was crystalized in the 1940s by mathematicians and engineers such as Claude E. Shannon and Norbert Wiener.[2] It became the backbone of the new transdisciplinary science of cybernetics, as well as making possible the computer revolution. Today, philosophy of information has become a significant area of research and development, with international societies and annual conferences devoted to the topic. It is typically understood as a relatively recent development: Luciano Floridi, beginning in the mid-1990s, has worked to establish it as a distinct area of philosophy.[3] From the beginning of the 1980s, however, Wu Kun had begun developing the area in China.[4] What remains largely overlooked are Ruyer's prescient, profound, and wide-ranging explorations of the philosophical implications of information theory. His work began significantly earlier, with two articles published in 1952: "Le problème de l'information et la cybernétique" ["The Problem of Information and Cybernetics"] and "La cybernétique, mythes et réalités" ["Cybernetics: Myths and Realities"]. This material was then much expanded into the first edition of this book, *La cybernétique et l'origine de l'information* [*Cybernetics and the Origin of Information*], which appeared in 1954 and was substantially revised for a second edition in 1967. Ruyer published another book that substantially contributed to many of the topics treated here, *Paradoxes de la conscience et limites de l'automatisme* [*The Paradoxes of Consciousness and the Limits of Automation*, 1966), raised these topics in many of his other books, and continued to write articles further developing his views on cybernetics, informatics, and the concept of information.[5]

Many good summary introductions to Ruyer and his philosophy are now available in English. I will not reproduce these basic considerations here, but simply refer the reader to these works.[6] I will restrict myself here to introducing this book by surveying of some of its most important arguments and giving some preliminary consideration to their continued relevance today. Let us begin with a brief characterization of Ruyer's position in and contribution to the philosophy of information.

Since their inception, cybernetics and information theory gained a polarized reception in France. Quite schematically, we can say that they were enthusiastically embraced by structuralists (such as Lévi-Strauss and Barthes), and critically dismissed by phenomenologists (such as Merleau-Ponty, and—outside France but deeply influential there—Heidegger). Later, poststructuralists such as Lyotard, Baudrillard, Deleuze, and Guattari combined aspects of these early receptions in more nuanced ways, but in general, remained quite critical of the notion of information.[7] Ruyer may be positioned as taking an early nuanced approach: he was critical of cybernetics and information theory in their then current and dominant forms, but saw in them much promise, and sought to reform and supplement them. Significantly, he was far more sanguine about the value and potential of the concept of information than most of his philosophical compatriots. In this, he is perhaps closest to that other outlier of French philosophy who has recently received renewed attention, Gilbert Simondon. Indeed, while *Cybernetics and the Origin of Information* seemed to achieve little reception, it was in fact a decisive, if almost entirely submerged, source of inspiration for Simondon's own profound reception of cybernetics and information theory.[8]

Tano S. Posteraro and Jon Roffe explain that

> Ruyer's primary mode of argumentation is the *reductio ad absurdum*, and the primary object of this *reductio* is the claim that any real being can be properly understood as an accretion of discrete parts organized *partes extra partes* in accordance with a fixed structure that transcends it. In short, his approach is to press this position to its limits to show how it both fails to account for basic phenomena . . . and that it ends up pointing to a remainder that it cannot explain.[9]

This is precisely the approach Ruyer takes to the mechanical theory of information, and its application in cybernetics. Ruyer aims to show that the postulates of cybernetics are absurd because they entail contradictions, and also to theorize a remainder by emphasizing the meaningful and creative aspects of information, which he believes must supplement the mechanical theory.

In sum, Ruyer's many arguments come down to this: information cannot simply be understood as a fixed pattern or structure that is communicated from a sender to a receiver and has its effects by the receiver simply reproducing this structure (as the mechanical model has it). Rather, Ruyer insists that both the sender and receiver must be capable of expression and interpretation, which are not akin to taking on an already formed form, but are the creation of form in a manner analogous to improvising on a given theme. This is intuitively the way that communication works in conscious language users, but Ruyer insists that other instances of morphogenesis, such as biological reproduction, also work in this way. Information, for Ruyer, is not then simply structure or pattern, but the *process* by which things are structured or patterned; it is not simply form, but *that which in-forms forms*. This view leads him to criticize the sufficiency of the mechanical theory of information, not in order simply to reject it, but to revise it by adding a further dimension. In the terms of Ruyer's own metaphysics, information must be accounted for in terms of the ability of consciousness to participate in a trans-spatial world of forms, essences, values, or ideas. Yet information also requires an actual (physical, spatial) world to inform, and so Ruyer insists that information must have a *mixed* or *dual* origin and nature.

THE CRITIQUE OF CYBERNETICS

Cybernetics, as summarily defined in the title of Wiener's general book on the topic, is the science of "communication and control in animals and machines."[10] Its originality lay in understanding machines and living beings according to the same model, with the mechanical theory of information acting as the "common measure" binding together these apparently different phenomena. What they have in common, supposedly, is the use of information in communication and control processes. Ruyer contests this identification, arguing that it remains essentially mechanistic, and cannot account for many important aspects of living beings. In Ruyer's words, cybernetics is a "false rapprochement"[11] of the living being and the machine, a rapprochement supposedly made possible by information theory. Concomitantly, he argues that the cybernetic theory of information cannot account for the "usual," psychological and linguistic sense of information, which involves *meaning*.

Ruyer notes that the mechanical theory of information was prepared for by theories such as pragmatism and behaviorism, which displace the idea of communication as depending on an interpretation of a meaningful sense with an emphasis on the effectuation of an action. Communication may then be understood in terms of the transmission of a message, understood as a *pattern* of some kind, which produces an *action*.[12] Information is understood as this message, and is defined as a pattern capable of producing an action—it then appears to be reducible to an observable, physical phenomenon. Understood in this way, information is quantifiable, and amenable to mathematical measurement and calculation.

Mechanical information is measured as a logarithm of probability, following the same formula as is used in thermodynamics to measure entropy, but with a reversed sign. There are some *prima facie* conceptual similarities between thermodynamics and information theory, insofar as information might be understood as a kind of order (called negentropy), and the absence or degradation of information, disorder (entropy). However, the extent to which this analogy can and should be pushed has been much debated: Wiener, the leading figure of cybernetics, tended to push it quite far, and this issue is precisely one Ruyer exploits in order to demonstrate the limits of mechanical information.

Critiques of information and cybernetics from a phenomenological perspective (for example, Heidegger, Merleau-Ponty) are relatively well known, but Ruyer departs from these in finding their construal of meaning in terms of consciousness or thought understood as a free and empty transcendence inadequate to account for the facts. He accords the cybernetic theory of information much more value than do the phenomenologists, on two counts: first, it poses problems in a very clear way, indicating with great perspicacity what mechanism *cannot* explain. Second, he believes that it correctly accounts for a necessary dimension of information, the physical dimension of communication.

Moreover, Ruyer's critique of cybernetics is resolutely not technophobic, as one can easily suspect with Heidegger and other critics. He is adamant that information machines will liberate humanity.[13] Moreover, cybernetic automata can help us understand the organic things they model and can thus advance our theoretical understanding in areas such as physiology and psychology.[14] It is, rather, the *postulates* of cybernetics that he critiques; in particular, the adequacy of the cybernetic proposal that living beings (and human beings in particular) are just organic information machines. These "mechanist postulates," Ruyer argues, lead cybernetics to a number of significant failures: the failure to understand the origin of information; the implicit admission of an impossible type of perpetual motion, the failure to understand meaning or sense; the failure to understand the perception of universals; and the failure to understand learning.[15] Despite dedicating much of his time to demonstrating these failures implied by the mechanist postulates, Ruyer also has a positive aim: to construct a less error-prone cybernetics, which would essentially acknowledge the limits of the mechanist postulates, and supplement what remains of cybernetics with his own metaphysics. This conception of cybernetics would understand machines as subordinated to, and "framed" by, the nervous systems of living beings, and at a higher and more essential level, the consciousness this implies. As for information itself, the positive conception would recognize consciousness and meaning as essential, and the mechanistic aspects as auxiliary (in the sense of a supplement and aid). While as we have just noted Ruyer elaborates on quite a number of errors of cybernetics, we will focus here on the central issue of Ruyer's book, which may also be seen as central to all of his claims, that of the *origin* of information.

A key to this issue is Wiener's admission that a mechanical operation can never create or augment information: machines can only conserve information by storing and transmitting it but cannot create or recreate it, and in fact (analogous to the second law of thermodynamics on which it is modeled) tend to degrade it. This allows Ruyer, quoting Wiener, to pose the problem of information's origin as follows:

> If "no operation by a machine on a message can gain information," and if, on the other hand, "there is no reason . . . why the essential mode of functioning of the living organism should not be the same as that of the automaton," then where does information come from?[16]

From a common sense perspective, we tend to think of information and communication as taking place between two conscious "centers," using technological media simply as a means. However, the cybernetic model suggests that these conscious centers (living organisms) are themselves no more than machines. So cybernetics proposes communication between machines but also insists that machines cannot create information. So, Ruyer reasonably asks, where then could information come from?

Ruyer draws out the inability of cybernetics to answer this question of origin in a number of ways. The first main argument he presents in *Cybernetics and the Origin of Information* draws out the analogy between thermodynamics and information theory to demonstrate that the mechanical postulates of cybernetics imply belief in the possibility of a "perpetual motion" in information machines, which is, however, as untenable as it is in other types of machines.[17] Following Wiener, Ruyer distinguishes three types of machines: *simple machines*, such as a clock, transform potential energy into kinetic energy. *Motor machines*, such as the steam engine, transform chemical energy into kinetic energy. Finally, *information machines* receive and transform information, the messages of communication and control. Ruyer suggests that simple machines and motor machines, both concerned with the power of movement, are like bodies without a head, whereas information machines are like the head (and nervous system).[18] The impossibility of building a perpetual motion machine—one which, once started, would not stop—is well known, and Ruyer develops this to point out the absurdity of a perpetual motion machine of the "third kind," that is, an information machine which would perpetually send and receive information on its own, without new input.[19]

In common experience, we tend to suppose that information machines such as the telephone require a conscious human sender and receiver. Yet in supposing that human beings *are* essentially themselves information machines, Ruyer argues, cybernetics is really supposing just such a perpetual motion of information machines, without recourse to an independent source of information. Information, he writes, is like the fuel or "supply" of information machines, and just as simple machines like the clock need someone or something to wind them up; just as motor machines such as the steam engine need new fuel to burn, so too do information machines require new inputs of information in order to function. Ruyer invites us to see the absurdity of a "perpetual motion" information machine as follows: "A telephone receiver," he writes, "can no more start to talk on its own than a wheel could start to move on its own simply because it has been attached to an axle."[20]

Ruyer concedes that the conservation of information in information machines can in principle be much more efficient than in other kinds of machines, and we cannot simply apply the analogy of Carnot's principle to them: noise that interferes with information does not degrade the message as surely as thermodynamic entropy degrades thermal energy.[21] Information, he suggests, can in principle be transmitted with 100 percent accuracy, and can be perfectly reproduced through techniques such as amplification and technical error correction; in this way an output can be extended indefinitely. But an information machine cannot *create* information, which would be required in order for the perpetual motion of the "third kind" envisioned above to function. Because of this, information machines "can no more freely create information than a simple machine can freely create work."[22] Ruyer calls this "the principle of the conservation of information," according to which there is never more information in the output of an information machine than in the input.[23]

This argument about perpetual motion considers information machines as *closed* systems. In chapter 5 of *Cybernetics and the Origin of Information*, Ruyer extends his argument to consider the more complex position, suggested by various cyberneticians, that they should instead be considered as *open* systems. The physics of thermodynamics has sought to explain how order can emerge despite entropy by suggesting that local order can be increased at the cost of a global increase of disorder, that is, an *overall* increase of entropy. This supposes open systems coupled with each other, exchanging energy. This model would then supply an answer to the question of the origin of information by again drawing out the analogy between information theory and the physics of (neg)entropy, understanding information as *order*. Order could be created by paying the "cost" of a greater disorder, and in coupled systems, the origin of information could be explained by a combination of random physical fluctuations, and the way they become ordered in coupled systems. Once again Ruyer considers this entirely inadequate, which he seeks to demonstarte by pointing to an equivocation in the meaning of the word "order," and the limits of applying the thermodynamic model to information. While physics can explain *homogenous* order, he argues, it cannot explain *complex structural order*, which requires an explanation in terms of life and consciousness, not simply physical negentropy. Even in a coupled system, the idea that complex structural order could arise spontaneously from random fluctuations seems to Ruyer as miraculous as the idea that signals conveying meaningful speech could spontaneously arise from pure noise on a telephone line.

Returning to this example of the telephone, then, Ruyer insists that the mechanistic explanation of information communication is completely implausible, and instead insists on an account that must appeal to consciousness:

> Psychological invention . . . goes from meaningful theme to meaningful theme. The man who improvises a message on the telephone first has a general idea of what he wants to communicate; this general theme evokes linguistic habits that are themselves abstract, which control the phonetic effectors and the specialised memories of the vocabulary.[24]

This now leads us from Ruyer's critique of cybernetics' inability to explain the origin of information to his own positive account of this origin.

THE ORIGIN OF INFORMATION

As already intimated, Ruyer asserts that what is missing from the cybernetic theory of information is a meaningful sense, expressed by the sender and understood by the receiver. As we have just seen, Ruyer locates this capacity in consciousness, and the "general idea" of what one wishes to communicate before the structured pattern that allows communication to take place is fully actualized. This "general idea" points to a necessary origin of information in a field of *possibilities*, which must be posited beyond the actualized space-time of physical data. The meaningful content of a message has a thematic character, meaning that it is composed of suggestions and possibilities, which are completed by consciousness. Ruyer argues that meaningful understanding requires a dimension of invention or creation, which can itself be characterized as a *recovery of information*. It is like the completion of a crossword puzzle, where partial empirically realized data are given, and access to a trans-spatial realm of themes suggests completed answers. Ruyer asserts that this is shown by the fact that machines are very limited in their ability to complete or reconstitute degraded messages, which human beings have no trouble finding meaning in: consciousness has access to the themes implied by partial data, while physical mechanisms do not.[25]

For Ruyer, then, the *reductio ad absurdum* arguments he has provided against the purely mechanistic postulates of cybernetics point to the necessity of a trans-spatial dimension with all the qualities he associates with consciousness: finality, absolute survey, form, idea, essence, value, meaning, and sense.[26] Ruyer also insists, however, that it is not consciousness as such, or the human brain, that is the origin of information; it is only a *medium* through which there is access to the trans-spatial world of thematic forms, which he understands on a broadly Platonic model.[27] He summarizes his conclusion as follows:

> The intuition of possibilities is the key to the problem of the origin of information. But this intuition is characteristic of consciousness and its relation with a "trans-spatial."[28]

However, Ruyer does not then argue for an exclusively idealist explanation of information, nor for a complete rejection of cybernetics. Rather, he argues for a mixed, or dual (*mixte*) origin of information. This origin must be located in "possible" trans-spatial themes, but they must be "actualized" in space-time, that is, mechanistically. The mechanistic aspects of information, identified by cybernetics, are only an auxiliary supplement and aid, but they are nevertheless a necessary one. Ruyer provides a number of examples, drawn from several fields, of how the "horizontal" space-time dimension of actualized information must work in conjunction with the "vertical" dimension of trans-spatial forms. I will briefly outline two of these: biological reproduction and language.

Biological reproduction is a major topic running throughout much of Ruyer's work, and the experimental embryology of his time provided one of the most significant models for his metaphysics of morphogenesis. This issue is integrally connected to information, because genes have frequently been understood on an informational model (especially since the discovery of DNA, in which genetic information has been thought to be "encoded"): species reproduce by passing information from one generation to another. On this model genes themselves are understood as informers, storing and communicating information in the form of instructions for creating a living being, information which is then "released" when reproduction takes place. The development of the living being would be a "structural amplification" of the information contained in the genes. This model involves, once again, a mechanical model of information, the sufficiency of which Ruyer here, as elsewhere, challenges.

For Ruyer, the theory of genes as informers is not sufficient because the reproduction of metazoans *"poses other enigmas than that of the duplication of genes."*[29] According to him,

> [t]he experimental study of development has demonstrated that the embryo develops in a thematic manner, and it is, because of this, capable of regulation in the manner of the intelligent interpreter capable of interpreting some general orders. It is never similar to a ribbon of the receiving apparatus on which the telegram is imprinted letter by letter.[30]

In embryonic development, the "informed" biological material does not simply passively receive the form it takes, but actively participates in the creation of form in a process that is comparable to improvisation on a theme. Development takes place progressively and in a variable—not an inflexibly predetermined—manner, and biological material "informed" by the same genetic information can develop quite differently under different conditions. The embryo contains an equipotentiality that allows its developing parts to take on quite different forms—to develop into different organs and limbs starting from primordia, for example—under different conditions, and this does not seem to be controlled by any direct genetic information. For Ruyer, genetic information "modulates" or "agitates," so has some important influence and part to play in reproduction, but a large part of the process involves an activity on the part of the informed biological material, which seems analogous to the comprehension of a meaningful theme by a consciousness. As is the case with psychological information, the receiver invents, as much as receives, the message. What informs the embryo, according to Ruyer, is not then the genes transmitted from the adult, but a trans-physical potential in which both adult and embryo participate. What this means for the informational theory of biological reproduction is that

> [t]he genes are not "informers," but modulators, or accidental agitators, of information. And above all, the primary information that modulates and agitates the genes has nothing to do with the mechanical communication which is all the cyberneticians envisage. It resembles rather information in the usual, spiritual or psychic sense of the term, inseparable from the apperception of a meaning.[31]

In fact, while Ruyer presents biological reproduction as a mixed case, it is also a limit case, in which the physical communication is at a minimum, almost negligible, and thematic participation is at a maximum. He emphasizes: "The individuals of a given species do not communicate between themselves in development. They develop themselves, they inform themselves, by direct participation in a specific potential."[32] However, he concedes that it is necessary to posit a minimal degree of communication between adult and embryo in order to explain hereditary traits, and the fact that individuals frequently resemble their parents more than they do other members of the species. Moreover, mutated or damaged genes reproduce their damages, therefore reproduction cannot be *pure* participation in a potential. The physical information transmitted by genes plays a role in "modulating" the biological development of an individual, but only—contrary to mainstream genetics—a minor and auxiliary one. On the basis of these arguments, Ruyer then concludes,

> From a certain point of view, then, the overall reproduction of a metazoan is a *mixture*. On the one hand, it is participation in a potential; on the other—since the genes transmitted, without being informers, have a disturbing influence—it is the result of a certain communication, which is probably in part mechanical.[33]

While the topic of biological reproduction is central to much of Ruyer's philosophy, as he himself notes language is a more telling example because "the definition of language is practically mixed up with the definition of information."[34] That is, we have a common understanding of what information means, tied to our experience of language. According to this common meaning and experience, as we have already noted, there is no information in a language without an understanding of it. But language also easily demonstrates that the "horizontal" aspect of information is present and necessary, in its physical dimensions (voice, writing, etc.) and in a code, the conventional system which structures it and allows the formation of patterns. Ruyer sums up his views as follows:

> A language always implies a set of mechanical and physiological media of informing communication, and a transmission of patterns in the spatiotemporal plane, which we can call "horizontal"; and it implies on the other hand two centres A and B, emitters and receivers (most often with a reversibility of these roles), capable of expression and of comprehension, that is, capable of participation which is this time "vertical," trans-physical, with a world of ideas, capable of converting the ideas into patterns, and the patterns into ideas. Moreover, language proper implies a code, more or less embodied in habits and memories; that is, a set of movements and conventional channellings, guiding the vertical participation and facilitating invention.[35]

Characteristically, Ruyer insists that the thematic, vertical dimension takes priority, and corresponds to a "primary information," the psychological sense of information as the understanding of meaning. Without this dimension, codes, which instantiate and conventionalize meanings, and facilitate their communication, could not be formed. Yet he equally insists that, despite its only-auxiliary nature, the coded dimension of language—conventional, physical patterns in space-time, which the cybernetic theory exclusively treats, is also necessary: "Consciousness does not read things without signs."[36]

In sum, Ruyer concludes that "[i]nformation is also a creation"[37] and is not simply the mechanical transmission of a given pattern, or the participation in trans-spatial ideas, but a mixture of the two. It is only by appreciating this double aspect, he believes, that the question of the origin of information can be correctly answered, the limitations of the mechanistic postulates of cybernetics be overcome, and cybernetics and information theory themselves be placed on a surer footing.

READING RUYER TODAY

Now that we have introduced in outline some of Ruyer's key arguments in *Cybernetics and the Origin of Information*, we can turn to the question of how we should approach the book as twenty-first century readers. For a start, this book is an important historical document, which alone makes the work significant given the good deal of scholarly interest in the history of cybernetics and its reception.[38] Such a historical reading of the book is relatively straightforward, whatever the challenges of interpretation, and needs little commentary or justification. But what of its philosophical content? How are we to weigh the validity and importance of Ruyer's seminal work on cybernetics and information today? It is of course a banality to observe that information technologies have come a very long way since Ruyer's main writings on this topic appeared at the beginning of the computer revolution. The obvious question which arises, then, is how Ruyer's philosophical positions hold up in light of these developments. A thorough investigation of this issue is well beyond the scope of this brief introductory survey. In the surveying spirit, however, I would like to present a few possible considerations toward answering this question.

Because of Ruyer's method of beginning with detailed examinations of scientific theories and technologies, this book does inevitably have a quite dated aspect. One case in point is machine translation, which Ruyer spends a good amount of time discussing in chapter 10. With the help of innovations such as deep learning, of which Ruyer could not have been aware, we have made a great deal of progress in this area. Today we have technologies (the most topical, at the time of this writing, being ChatGPT) that can, to an impressive extent, imitate the contextual semantic interpretation that Ruyer insists provides a hurdle to machine translation, and caused the researchers of his time to despair. This "datedness" is to some extent mitigated, however, by the fact that Ruyer's main point, with all his arguments in this book, is not to argue for any "in principle" limits of information technologies in terms of the *results* they will be able to achieve. He is perfectly willing to admit that machines can be built that can and will be able to imitate forms of human reasoning better than human beings are capable of. Yet the key word here is *imitate*, rather than *model*. Ruyer's main contention is that cybernetics is wrong in believing that mechanical systems can operate as models of how human beings and other living, conscious organisms actually think and function. It is the processes, not the results, on which his claims rest.[39]

To see this clearly, we can take another case in point, his discussion of computer chess. Once again experiments in this area were still in their infancy when Ruyer was writing this book, and he points to the difficulties computers have playing the game because of the sheer number of possible moves that it would seem to have to calculate. Yet in 1997 the computer Deep Blue defeated chess world champion Garry Kasparov, and today it is commonplace that computers have the advantage over human players. While a cursory reading might suggest that in hindsight Ruyer was simply wrong here, the principles on which he argued remain correct. Chess is still an "unsolvable" game, in the sense that the best possible move in any position cannot be calculated because there are simply too many possibilities, even for today's best computer processing. Instead, the success of computer chess came through strategies that acknowledged this limitation. Deep Blue, an expert system, calculated the best option from a finite (though very large) number of possible moves, enough to defeat a human opponent. Research that led to this point acknowledged that this "raw calculation" is not the way that humans in fact play chess. Roughly, humans tend to give deeper consideration of fewer possibilities, while the strategy Deep Blue employed gives shallow consideration to a greater number of possibilities. Today, computers trained with machine learning methods can easily defeat the best human players, but one area of research concerns whether human-assisted machines might be better at chess than a machine alone, again accenting the difference in game play of each. To the extent that Ruyer's point is that humans play chess (and other games) in a way that is not raw calculation, then this point remains valid. This of course is not to say that all Ruyer's arguments are convincing, and I cannot spare the reader from doing the work of assessment themselves.

Rather than dwell further on the successes or failures of Ruyer's specific arguments in light of current technologies (which would be an extensive task), I want to point to what I believe is the deeper and more persistent relevance of

Ruyer's work on philosophical questions concerning information itself. To do this, I will take the expedient of referring to Floridi's list of "Open Problems in the Philosophy of Information."[40] Taking Floridi as a guide, we can see that many of the problems Ruyer worked on are far from having yet been solved. Floridi identifies eighteen "open problems," organized into five groups: the analysis of information, semantics, intelligence, nature, and values. Many of the questions are interconnected, and Ruyer's work conceivably intersects with a large number of them. However, we can single out a few as being of particular relevance. Problem 1 is "The Elementary Problem: What Is Information?" It is the central issue in the philosophy of information, and all Ruyer's work on information theory, including his critique of the sufficiency of the mechanistic model and his conception of information as requiring a dual nature and origin, contribute to addressing it.

Ruyer's work is highly relevant for the area of semantics, and in particular bears on Problem 7, "Informational Semantics: Can Information Explain Meaning?," which includes the question "Can semantic phenomena be explained as aspects of the empirical world?"[41] The question of the relation of quantifiable information to semantic information remains a prime issue of philosophical debate. As we have seen, Ruyer argued extensively that the cybernetic theory of information cannot explain meaning, and a theory of psychological information, along with the relations between these two, is required. Mark B. N. Hansen—in addition to drawing out other interesting applications of Ruyer's philosophy for issues in new media, such as the relevance of absolute survey for understanding virtual reality—has suggested that Ruyer's arguments about information are broadly consistent with those of Donald MacKay, a cyberneticist who argued for the necessity of expanding information theory to account for the semantic dimension.[42]

The distinction between mechanical information and signification (psychological, meaningful information) that Ruyer insists is essential,[43] but is obfuscated in much of the cybernetic literature, is posed with much greater clarity today in philosophy of information in terms of the data/semantics distinction. Data refer to the syntactic or physical pattern, while semantics refers to the additional dimension of the *meaningful interpretation* of data.[44] This certainly does not mean that problems of the relationship between data and semantics have been solved, but rather that they can be posed more adroitly, and that there is less need today to labor the distinctions that Ruyer does.

These issues of data and semantics are related to the next group of problems, intelligence, especially concerning the possibility of artificial intelligence. As far as research into "thinking machines" is concerned, cybernetics largely gave way to the paradigm of artificial intelligence (AI) by the 1970s, but many of the philosophical issues and problems that Ruyer interrogated in the former persist in the latter, up to the present day.[45] Again, insofar as

Ruyer's work challenged the cybernetic modeling of minds and machines according to the same principles, it has the potential for intervening in debates around AI, which have today become more pressing than ever. The difference between the mechanical model and the semantic (thematic, in Ruyer's terms) model of information lies at the heart of claims regarding artificial intelligence. Ruyer's arguments against the sufficiency of *functioning* to explain psychological information seem confluent with arguments against functionalism in analytic philosophy of mind, and related issues in artificial intelligence, such as John R. Searle's "Chinese room" or Ned Block's "China brain" arguments.[46] While contemporary media is full of reports of great advances being made toward true (or "strong") artificial intelligence, there remains a wide abyss between the hopes and hypes of tech investors, and the far less sanguine assessments of a majority of philosophers on the topic. Here a cursory reading of some of Ruyer's claims—such as the inability of machines to correct mistakes (or to reconstitute degraded information, as noted above)—might suggest an obsolescence in light of more recent technological developments (such as autocorrect in word processing applications), but a closer reading reveals that his various qualifications showed a good understanding of what information machines would and would not be capable of, in principle (he concedes that many kinds of limited mechanical correction or reconstitution are possible, but they remain extrapolations from given data, and can never be a true invention of information, as consciousness is capable of.)[47]

The metaphysical character of Ruyer's philosophical work lends itself to contributions in the next set of problems Floridi specifies, those concerning its nature. The first is Problem 15: "Wiener's Problem: What Is the Ontological Status of Information?" According to Wiener, information is neither matter nor energy, but belongs in a unique category of its own, and this opens questions regarding the fundamental ontology of information in relation to traditional philosophical classifications. Today, it remains an open question whether information is physical, ideal, or something else entirely. Ruyer's argument for the mixed character of information, as both ideal and physical, has an obvious bearing here.

Problem 16 is "The Problem of Localization: Can Information Be Naturalized?," and Floridi poses this issue in terms which readily allow us to see Ruyer's relevance:

> The location of information is related to the question whether there can be information without an informee, or whether information, in at least some crucial sense of the word, is essentially parasitic on the semantics in the mind of the informee.[48]

This recalls Ruyer's central question of the *origin* of information. And in surveying the range of possible answers to the question of information's localization, Floridi writes, "Or could it even be elsewhere, in a third world, intellectually accessible by intelligent beings but not ontologically dependent on them (Platonism)?"[49]

We could of course insert into this alternative Ruyer's claims about the origin of meaning in absolute forms, participation in which gives meaningful information, itself necessary for information in the mechanical, cybernetic sense (with the caveat that for him this second dimension is *also* crucial). In short, Ruyer's arguments bear strongly on questions of localization insofar as he suggests that cybernetic information concerns the horizontal world of space-time, while meaning requires reference to a vertical world of transcendent forms.

In addition to the problems that Floridi identifies, Ruyer's work has continued relevance for other problems, such as the role of information in biology.[50] Georges Chapouthier has argued for the "modernity" of Ruyer's work in this area, primarily in resisting what he considers to be one of the most egregious and pervasive errors in twentieth-century biology, the identification of the three concepts of negentropy, order, and information.[51] He cites Léon Brillioun in particular as responsible for this confusion. It is an error because negentropy is a *quantitative* concept, while order and information are *qualitative* concepts. Chapouthier claims that only in rare cases can order be aligned with a quantitative measurement, and in most cases, there is nothing that allows us to measure whether one system is more ordered than another. Moreover, as we have already seen, the common sense of information as a "unit of knowledge"—or its semantic meaning—is irreducible to the quantitative measure of Shannon's theory. As the reader of this book will see, this distinction between the quantitative, measurable notion of information as "negentropy," and the qualitative notions of order and conscious, meaningful information are crucial to many of Ruyer's arguments (see in particular chapters 5 and 6). While Chapouthier concedes that Ruyer's metaphysical dualism is unlikely to find favor among today's scientists (and it is not a position he himself subscribes to), it nevertheless has a strategic value in resisting the excessive reductionism of some scientific theories, and in particular the erroneous identification of negentropy, order, and information, by pointing to their problematic philosophical implications.

I would add here that we can generalize Chapouthier's last point: while Ruyer's critiques and supplements of information theory neatly conform to his broader metaphysics, they may to a large extent be appreciated independently. One does not need to buy into his neofinalism, reversed epiphenomenalism, or metaphysics of trans-spatial forms in order to recognize and appreciate the important early contributions Ruyer made to the philosophy of information. As we have seen, these contributions point to the limits of mechanistic information theory, and outline in what ways it needs to be supplemented in order to fulfill its ambition and potential. We might potentially agree with these limits and need for supplementation, but nevertheless remain skeptical that these supplements must conform to the metaphysical models Ruyer suggests.[52]

Reading Ruyer today requires a lot of careful thought and reflection in order to fully appreciate his arguments, both in their historical context and in light of contemporary technological and theoretical developments. There seems to be diminishing time and space for such patient work in our current environment of demand for immediate results. But this patient, reflective work is a necessary resistance to this accelerated world that the very information technologies Ruyer interrogates here have helped to bring about, and essential if we are to learn to live with them in a way as beneficial as Ruyer suggests we might. As I hope the above, brief and provisional points are enough to indicate, many of the questions in philosophy of information interrogated by Ruyer, well over a half-century ago, have since received no definitive philosophical answer, and remain open. *Cybernetics and the Origin of Information* therefore cannot be considered as a book firmly closed by the march of time. On the contrary, the information revolution and the host of increasingly pressing problems to which it has given rise invites us to reopen this rich and fascinating work.

BIBLIOGRAPHY

Barthélémy, Jean-Hugues. *Penser l'individuation: Simondon et la philosophie de la nature*. Paris: L'Harmattan, 2005.

Block, Ned. "Troubles with Functionalism." *Minnesota Studies in the Philosophy of Science* 9 (1978): 261–325.

Chapouthier, Georges. "Information, structure et forme dans la pensée de Raymond Ruyer," *Revue Philosophique de la France et de l'Étranger* 203, no. 1 (2013): 21–28.

Dupuy, Jean-Pierre. *On the Origins of Cognitive Science: The Mechanization of the Mind*. Trans. M. B. DeBevoise. Cambridge, Massachusetts: MIT Press, 2009.

Floridi, Luciano. *The Philosophy of Information*. Oxford: Oxford University Press, 2011.

———. "Semantic Conceptions of Information," *Stanford Encyclopedia of Philosophy* https://plato.stanford.edu/entries/information-semantic/ (revised January 7, 2015).

Gagnon, Philippe. *La réalité du champ axiologique: Cybernétique et pensée de l'information chez Raymond Ruyer*. Louvain-la-Neuve: Les Éditions Chromatika, 2018.

Geoghegan, Bernard Dionysius. *Code: From Information Theory to French Theory*. Durham and London: Duke University Press, 2023.

Hansen, Mark B. N. *New Philosophy for New Media*. New York: MIT Press, 2004.
Hayles, N. Katherine. *How We Became Posthuman: Virtual Bodies in Cybernetics, Literature, and Informatics*. Chicago: University of Chicago Press, 1999.
Iliadis, Andrew. "Mechanology: Machine Typologies and the Birth of Philosophy of Technology in France (1932–1958)," *Systema* 3, no. 1 (2015): 131–44.
———. "Introduction to Ontologies of Difference: The Philosophies of Gilbert Simondon and Raymond Ruyer," *Deleuze Studies* 11, no. 4 (2017): 491–97.
Lafontaine, Céline. "The Cybernetic Matrix of 'French Theory,'" *Theory, Culture & Society* 24, no. 5 (2007): 27–46.
Li, Guowu. "Information Philosophy in China: Professor Wu Kun's 30 Years of Academic Thinking in Information Philosophy," *TripleC* 9, no. 2 (2011): 316–21.
Pickering, Andrew. *The Cybernetic Brain: Sketches of Another Future*. Chicago: University of Chicago Press, 2011.
Ruyer, Raymond. "Le problème de l'information et la cybernétique." *Journal de psychologie normale et pathologique* 45 (1952): 385–418.
———. "Les informations de présence." *Revue Philosophique de la France et de l'Étranger* 152 (1962): 197–218.
———. "Quasi-information, psychologisme et culturalisme," *Revue de métaphysique et de morale* 70, no. 4 (1965): 385–418.
———. "La quasi-information: Réflexions en marge de deux ouvrages récents." *Revue philosophique de la France et de l'Étranger* 155 (1965): 285–302.
———. "Cybernétique et informatique." In *La philosophie contemporaine, vol. II: Philosophie des sciences*. Ed. R. Klibansky. Florence: La Nuova Italia, 1966.
———. *Paradoxes de la conscience et limites de l'automatisme*. Paris: Albin Michel, 1966.
———. *La cybernétique et l'origine de l'information*. Paris: Flammarion, 1954 (2nd Rev. Ed. 1968).
———. "Raymond Ruyer par lui-même." *Études philosophiques* 80, no. 1 (2007): 3–14.
———. *Neofinalism*. Trans. Alyosha Edlebi. Minneapolis: University of Minnesota Press, 2016.
———. *The Genesis of Living Forms*. Trans. Jon Roffe and Nicholas B. de Weydenthal. London and New York: Rowman & Littlefield International, 2019.
Ruyer, Raymond, Tano S. Posteraro, and Jon Roffe. "Instinct, Consciousness, Life: Ruyer contra Bergson." *Angelaki* 24, no. 5 (2019): 124–47.
Searle, John. "Minds, Brains and Programs." *Behavioral and Brain Sciences* 3 (1980), 417–457.
Smith, Daniel W. "Raymond Ruyer and the Metaphysics of Absolute Forms." *Parrhesia* 27 (2017): 116–28.
Veilhan, Alix. "Raymond Ruyer et la cybernétique." *Philosophie* 149 (2021/2): 41–57.
Wiener, Norbert. *Cybernetics, or Control and Communication in the Animal and the Machine*. 2nd Ed. (1964). Cambridge: MIT Press, 1991.

Note on the Translation

There were two editions of Raymond Ruyer's book *La Cybernétique et l'origin de l'information*, both published by Flammarion. The first appeared in 1954, the second in 1967. There are some significant differences between the two editions, the second having been substantially revised. The most obvious difference is that the second edition adds a very long final chapter, "The Problems of Cybernetics in 1967," and removes the first edition's brief "Summary and Conclusion." The rest of the text of the second edition has also been substantially shortened, with some sections of chapters throughout the book removed. Often, this has involved sections with diagrams and examples that illustrate and expand on the argument but are not essential to it. And finally, some terminology has been altered, and other minor edits made here and there.

With the current translation we have combined both editions, in the spirit of Norman Kemp Smith's classic translation of Kant's *Critique of Pure Reason* (though fortunately Ruyer's texts presents a far less complex challenge). Instead of designating an A and B edition, however, we have adopted the simpler practice of, wherever possible, placing in bold square brackets—[ ]—the text of the first edition that has been removed from the second. We have indicated more complicated changes between the two editions with endnotes. Given this complication of the text, we have chosen to refrain from further complication and have not attempted to include original pagination in the margins, as has been the practice with previous books published in this series.

Square brackets not in bold—[ ]—indicates original French text.

An asterisk—*—designates a word in English (and generally italicized) in the original French text.

Endnotes we have added ourselves, which are not translations of the original text, are designated TN (Translators' Note).

As Jon Roffe and Nicholas B. de Weydenthal note with their translation of Ruyer's *The Genesis of Living Forms* (also published by Rowman & Littlefield in the Groundworks series), Ruyer's method of citation is "to be frank, fairly impressionistic" (p. viii). We have in many cases added fuller citation information, and where appropriate we have used the original texts quoted rather than (re)translate Ruyer's own renderings (which are often from English texts). We have also generally follwed existing English translations of French texts quoted.

TERMINOLOGY

Several translation choices deserve some comment.

Liaison. This term could be variously translated as "connection" or "bond," or left as "liaison" as the term has entered the English language. This is an important term in Ruyer's metaphysics and works in several registers, not all of which are captured by a single translation choice (including the English "liaison"). Alyosha Edlebi favors "bond" in his translation of *Neofinalism*. Due to the fact that it is most often machines under discussion here, we have chosen the term "connection" in most contexts. However, it needs to be borne in mind that Ruyer distinguishes a "primary" kind of connection, associated with consciousness, from the "secondary" one that applies to physically instantiated machines (see in particular the section "Connections and Consciousness" in chapter 6). In the discussion of atoms and molecules, however (also in chapter 6), we have chosen to translate *liaison* as "bond" given the conventional English use of this term in this context.

Technique. This has been variously translated as "technics," "technology," or "technique" depending on context. The French *la technique*, used to name a discipline or field, combines the senses of technology and technique, and—as has become familiar through translations of works by other French philosophers of technology, such as Gilbert Simondon and Bernard Stiegler—may be translated as "technics."

Sens and *signification*. Both these terms have usually been translated as "meaning." Edlebi prefers "sense" for *sens*, but in our view "meaning" is preferable due to its lower degree of ambiguity. "Signification" still tends to evoke a particular theory of semiotic or linguistic meaning popularized through French structuralism, which is not generally what Ruyer has in mind when he uses this term.

Montage; *monter*. Generally translated as "assembly" and "to set up," but with exceptions depending on context. "Assembly" has usually seemed better in technical contexts (e.g., the assembly of a machine), and "set up" in the context of the living organism, its nervous system, and consciousness, but these are often mixed tightly together in the text. It should be borne in mind that these English terms are translating a single French term.

Ébauche. Another essential term for Ruyer, which might be translated as "stub," "stem," or "sketch," but which has a specific technical English equivalent in the context of biology, which is "primordium." In such contexts we have used "primordium," but occasionally the more generic "sketch" has been preferred, and in one place we have left the French term untranslated. See chapter 10, note 50, for further explanation.

Tuyau. This has been variously translated as "tube" or "pipe," depending on context. In *The Genesis of Living Forms*, Roffe and de Weydenthal have chosen "duct" when the term refers to organic formations, but we have decided "tube" is sufficient here in such cases.

Introduction

Cybernetics (which comes from a Greek word meaning "to govern") can be defined as the science of control by[1] information machines, whether these machines are natural, such as organic machines, or artificial. Cybernetics began in America in the 1940s with the work of various mathematicians (Norbert Wiener, John von Neumann), physicists and engineers (Vannevar Bush, Julian Bigelow), and physiologists (Walter B. Cannon, Warren S. McCulloch). [In France, at roughly the same time, a similar convergence occurred between the physiologist Louis Lapicque and the engineer Louis P. Couffignal.]

Simple machines, without creating work, change the force/displacement relation. Clock mechanisms transform the energy of a spring into movement. Motor machines, such as the steam engine, transform chemical energy into kinetic energy. The most characteristic machines of the twentieth century—as opposed to the simple machines of the Greeks, or the clockwork mechanisms of the eighteenth-century, or the high-powered motor machines of the nineteenth century—are information machines.[2] High frequency transmitters, which are characteristic of much contemporary technology, are of little interest as machines of power. They are very inadequate machines for transmitting energy since they radiate it in all directions. They are, above all, machines for transmitting or receiving information.

Clearly there is no clockwork movement in an organic body, and the automata of the eighteenth century had only a superficial resemblance to living beings. However, organisms include both simple machines and motor machines. The human body contains numerous levers and is driven by the chemical energy of nutrients. Machines, apart from information machines, are like bodies without a head, and they can replace manual workers since all we need from them is their labor capacity. But once they are equipped with servomechanisms of information, and thus become capable of self-control, machines start to resemble complete organisms with a head, that is, with a nervous system and organs of perception. They can work toward a given end, despite accidental interferences. They can then replace intellectual workers, from whom we demand vigilance and initiative in the fulfilment of their duty.

According to most of the cyberneticians, the sense organs and nervous system of living beings are in principle nothing other than information machines and are controlled by information.

In *Erewhon* (1872), Samuel Butler suggested that machines represented a new realm that was dangerous to humans, and he foresaw, as a particularly critical point, the day when machines would become truly automatic—by which he meant, without using the word, capable of controlling themselves through information: "As yet, the machines receive their impression through the agency of man's senses: one traveling machine calls to another in a shrill accent of alarm and the other instantly retires; but it is through the ears of the driver that the voice of the one has acted upon the other. Had there been no driver, the callee would have been deaf to the caller. There was a time when it must have seemed highly improbable that machines should learn to make their wants known by sound, even through the ears of man; may we not conceive, then, that a day will come when those ears will be no longer needed, and the hearing will be done by the delicacy of the machine's own construction?"[3]

This day has come. Machines inform each other, and they inform themselves.

INFORMATION

The word "information," in its usual sense, seems necessarily to include an element of consciousness and meaning, which even seems essential. Most of us try to be well informed on politics, or on the progress of technology, for the pleasure of possessing this knowledge. Information, in the ordinary sense of the term, is the transmission to a conscious being of something meaningful, such as a concept, by means of a more or less conventional message and a spatiotemporal pattern*[4] (printed materials, telephone messages, sound waves, etc.). The apprehension of the meaning is the end, the communication of the pattern is the means. In some cases, we may need information when we have a utilitarian purpose in mind; information then becomes the means, and the action it triggers or controls becomes the end. Pragmatism and behaviorism long ago learned from psychologists to emphasize action rather than consciousness, and cybernetics has rigorously adopted this point of view. In information, meaning or consciousness is not essential; or more accurately, the meaning of information is nothing other than the set of actions it triggers and controls. If I say to a man who shares the same office with me, "It's too hot in here; let's open the window," and the man answers, "You're right; it's hot; hurry up and open it," there seems to have been an exchange of conscious impressions, even more obviously than there has been the preparation for a movement. Nonetheless, psychology—even classical and academic psychology—long ago recognized that a consciousness that causes no reaction can hardly be called a consciousness. I might be so absorbed in a specific task that I scarcely feel the excessive temperature; and consciousness appears only at the moment I respond. My body may have reacted well before my consciousness, through mechanisms of thermal regulation such as perspiration, which operate unconsciously. Similarly, when riding a bicycle or driving a car, I will brake when I suddenly confront an obstacle, even before I feel any fear. One might even say that "to see an obstacle" or "to be aware of an obstacle" means "to avoid the obstacle." If I am distracted, and I look in the general direction of the obstacle without responding, and then crash into it, can it be said that I *saw* it? If an automaton seated next to me, analogous to the artificial animals of William Grey Walter, detected the obstacle through its photo-electric cell and managed to avoid it, who would have given a better impression of being conscious, the automaton or me?

If the room in which I work was air-conditioned by means of reflex machines, a temperature device would have been informed of the temperature in the room and in turn would have informed the heating and ventilation devices. Between them, there was no exchange of impressions, yet the result would have been at least as effective as my conscious reactions. If "information," in the sense of a transmission from machine to machine, is metaphorical, it must be recognized that this cybernetic metaphor seems to contain virtually all the essentials of reality.

Any effective communication of a structure can thus, it seems, be called information. It would not be illegitimate to say that changes in barometric pressure "inform" the barometer, or that sound waves, electrically transmitted by telephone or radio, "inform" the receiving or recording devices. Moreover, this objective definition of information—which is also consistent with the original meaning of the word—has the great advantage of making it accessible to measurement. If information is essentially the progression of a structurally efficacious order, it will be the opposite of a "de-structuration," a breakdown, a decrease of order. This decrease of order has a name in physics: entropy. Information can thus be regarded as the opposite of entropy, and it will be measurable as this.[5]

THIS DEFINITION IS PARADOXICAL

The cybernetic conception of information is nonetheless paradoxical, despite all the goodwill involved in recognizing its elements of truth. In the transmission of a pattern[6] between two machines, or between parts of the same machine, a form winds up being transmitted as a significant unit because a conscious being can become aware of the end result *as* a form. But the transmission itself, if it is mechanical, is only the transmission of a pattern, a structural order without internal unity. A conscious being, in apprehending the pattern as a *whole*, turns it into a form; but under analysis, the transmission takes place in the machine through a step-by-step functioning, or by partial and isolated functionings. The sinuous line that I perceive as a whole was plotted out point by point, or section by section, by the pen of the barometer recorder. Similarly, the sound waves on the phone have been reconstructed by electrical relays, and if there were not an ear—or rather, a conscious "I"—listening at all levels of the information machine, we would never find a form, properly speaking, but only fragmented functionings. The use of the machine by a human being to get "information," in the psychological sense, deceives us about the nature of the machine. A formal order is attributed to every level of the machine even though this order only appears at the end, thanks to something that does not belong to the machine itself. If I forget to turn off my radio and the speaker recites a poem during my absence; and if, moreover, at the radio studio, the recorded tape is running without any supervision, there is obviously not a "recitation of a poem," but only uncoordinated elementary functionings that have a consistent structure only in a very precarious and residual manner. There is no "recited poem" here, any more than there is a "profile of Napoleon" on a rock that has been shaped by natural forces. If the physical world and the world of machines were left to themselves, everything would spontaneously fall into disorder, and it would become clear that there had never been a real or consistent order—in other words, that there had never been any information.

In certain intermediate mechanisms of information machines, there can be threshold or pivotal effects[7] that seem to bring about, in a practical manner, a *summation* or "consideration of the whole" that might seem to transform the structures and elementary functionings into a form or an authentic order. For example, the "automatic reader" created by Walter Pitts and Warren McCulloch, which allows blind people to listen to Braille, transposes the shape of the letters, which are detected by a photoelectric scanner, into sounds that listeners can learn and identify, just as they learned to identify the tactile sensations of the Braille letters.[8]

This device could quite possibly be improved using pivotal effects obtained either by a set of photoelectric cells arranged on the surface, or by a single electron tube with directed flow and a grid screen using surface structures, to make such a device really "read" a printed text. But by themselves, these threshold and pivotal effects are still a step-by-step functioning, and not information. An appropriate key opens a lock through a point-by-point correspondence of patterns[9] and not through a transmission of information.

To say that the door opens only if the lock "recognizes" the key is to make a metaphor of dubious interest. It is obvious that without the awareness of the blind person, McCulloch's machine is as useless as the recitation of a poem from a speaker in an empty room.

THE ORIGIN OF INFORMATION AND THE POSTULATES OF CYBERNETICS

The paradox of the *nature* of information is combined with another paradox of the *origin* of information. To my knowledge, cybernetics has never explicitly stated its views on the origin of information. Yet the paradox results from the merger of two theses set forth by Norbert Wiener. The first of these theses is that information machines cannot gain information: there is never more information in the message that comes out of a machine than in the message given to it. Practically, there is less, because of unavoidable effects that, according to the laws of thermodynamics, tend to increase entropy, disorganization, and disinformation. The second is that brains and nervous systems are information machines, certainly more sophisticated than industrially built machines, but of the same order as them: they do not contain any transcendent property nor are they impossible to imitate by a mechanism.

If we combine these two theses, it becomes impossible to understand what the origin of information might be. If nervous systems are information machines and nothing else, according to the second thesis, then we must be able to apply to them the "principle of the conservation of information" given in the first thesis. There is never more information in the "output" of a brain than in its "input." When I send a message, it is "I" who wrote it before entering it into the machine. To common sense, I am the origin of the information; the machine is a transmission channel. Common sense would probably not venture to add, given enough time for reflection, that the "I" is the absolute creator of information. It knows very well that the sent message is not a pure creation, even when the author did not use a manual of etiquette or a guide to commercial correspondence. But it also knows that galvanizing themes [*thèmes inspirateur*] have contributed to the elaboration of the message in a quite particular manner. The "I" is not an absolute origin, but neither is it a simple organ of transmission. In the elaboration of even the most unassuming message, one can clearly see that it is not simply a matter of allowing the brain to function; it is also about inserting into space (and giving to the machines functioning in that space) a "supply" [*aliment*] that cannot simply be taken from another part of space.

PERPETUAL MOTION OF THE THIRD KIND

If the cybernetic arguments against this impression were correct, then a "perpetual motion of the third kind" would be possible. Let us remember the three main kinds of machines we distinguished, following Norbert Weiner: simple machines with clockwork movements; motor machines with external energy sources, such as the steam engine; and information machines. A simple machine cannot create work for free, since a clockwork mechanism must be wound up by hand. Hence the impossibility of a perpetual motion of the first kind,[10] and the long-recognized chimerical nature of systems in which enough work could be created, simply through a new mode of assembly, to compensate for the unavoidable energy losses due to friction. A heat engine can only work with an external energy source, such as coal or gas. Moreover, according to Carnot's principle, it degrades this energy, since it requires two energy sources, at different temperatures, between which the energy used passes from less probable states (a temperature higher than that of the environment) to more probable states (a temperature identical to the environment). A ship on a tropical sea cannot cool the sea; its pistons must cool the steam first heated at great cost in its boilers. Hence the impossibility of perpetual motion of the second kind. Finally, information machines are analogous to both simple machines and heat engines: theoretically, they can at best retain the information they transmit, though practically the information is always degraded.

To be sure, as information machines, not energetic machines, their output is certainly much better than that of heat engines, and we cannot apply to them a principle analogous to Carnot's principle. They do not have a condensing unit like heat engines, in which information, after being processed, would come out degraded and close to an absolute zero degree of information. Theoretically, there is nothing that would prevent an output with 100 percent accuracy. On the one hand, one utilizes information when reading a message, but the information is not altered, or is altered only in an infinitesimal way; on the other hand, background noises or interferences disruptive of information can be reduced asymptotically or debugged mechanically when the elements of a message threaten to fall below a certain threshold of security. Messages in binary systems, using 0 and 1, and messages in Morse code, using lines and points, can be debugged in this way. Lines that are too short and points that are too long can be normalized through relays. If the signal falls below the threshold of security, but also above the threshold of the relay's functioning, it is possible that the error will be aggravated rather than corrected. But precise adjustments can avoid such accidents.

It is this clean, and even theoretically perfect output that allows a given piece of information to be extended indefinitely. Copies of a newspaper or a photograph can be multiplied almost indefinitely. A pattern*[11] of information can also be amplified. But reproducing or amplifying a pattern does not increase the information itself. If information machines escape Carnot's principle and its limit on output, they cannot escape the principle of the conservation of information. They can no more freely create information than a simple machine can freely create work.

If a phone line is too long, the structure of the sound waves becomes muddied, and the receiver can only pick up static. While suitably spaced relays can avoid this nuisance, no conceivable system can avoid putting a clearly formulated message into the line. A telephone receiver can no more start to talk on its own than a wheel could start to move on its own simply because it has been attached to an axle. Similarly, it would be impossible to send a telephone message by automatically sending an emission of "static" that would progressively transform itself into a message at the receiving end, just as it would be impossible to set a boat in motion on the sea by relying on the lucky coincidence that the water molecules striking the stern of the ship would do so at a speed that was constantly greater than those striking the bow. Strictly speaking, it would not be impossible for the static on a phone or radio to reestablish a local detail of information that had been previously lost in the background noise. Nor would it be impossible, in an electronic calculator, for an amount erroneously subtracted from a sum by a malfunctioning switch to be added by the malfunction of another switch in a subsequent stage, with the second error correcting the first—just as it would not be impossible for a microscopic particle to travel from A to B simply by relying on molecular agitation. But it would be unwise to rely on these kinds of fluctuations to produce a message or to travel across the ocean. Travel requires coal or oil. A ship equipped with highly sophisticated machinery, but without combustible fuel, is not enough. To send a message, an information machine, admirable though it may be, is not enough either. It needs human beings to feed it, that is, to provide it with messages to transmit. If these human beings were of the same type of machine as those they were feeding, if they could not create information, we do not understand how messages could be sent. Perpetual motion of the third kind is as impossible as perpetual motion of the first or second kind. What is it, in information machines, that plays the role of coal or gas in heat engines? The purpose of this book is to answer this question.

THE PRACTICAL INTEREST OF CYBERNETICS

It is only in appearance that this book has a negative and critical aspect. The object of our critique is the postulates of cybernetics, not cybernetics itself, whose practical and theoretical interest is immense. The academic fears about the automation of human beings by automata seem absurd to us. Information machines, servomechanisms, and automation of all kinds will liberate human beings, not only from manual work, but also from everything that is "slavish" in the work of surveillance or control. They will liberate human brains just as high-powered machines have begun to liberate human muscles. They will liberate everything by increasing human power. Edmund C. Berkeley was no doubt correct when he wrote that electronic machines will inaugurate a new era in human thinking in the same way that the tank inaugurated a new era in tactics: "In the Middle Ages, there were few kinds of weapons, and it was easy for a man to protect himself against most of them by wearing armor. As gunpowder came into use, a man could no longer carry the weight of the armor that would protect him, and so armor was given up. But in 1917, armor, equipped with a motor and carrying the man and his weapons, came back into service—as the tank."[12] Likewise, today, there is an imbalance between the naked brain of humans and their own science. The brain is too weak to bear the weight of the enormous amount of information accumulated in libraries through printing. Only motorized brains will be able to utilize this accumulated information and make it viable. The era of the intellectual foot soldier is about to come to an end.

Berkeley was thinking particularly about calculating machines. But industrial automata will help humans even more to bear the weight, not only of their accumulated information, but also of their accumulated technologies. Bergson, meditating on the accumulated weight of material technologies, wrote that "this enlarged body awaits a supplement of soul."[13] But this body of machines first waits to be perfected and neutralized by servomechanisms. It is then and only then that humans—and the human soul—will be freed from the mechanical body of civilization, whose functioning will become as unconscious as the physiological functioning of a healthy organism. In a civilization where machines are starting to reign, but where servomechanisms do not yet exist, humans must themselves play the role of the "serf," the servant of their machines. The harshest slavery, as Georges Friedmann noted, coincides with the beginning of automation, when the machine imposes its own pace on the worker.[14] Thanks to information machines, which have been added to power machines like a head to a body, the very brains of humans have finally been liberated. A single human brain, in relation to the incomplete automation machines that it has to operate, is as insufficient as the brain of the gigantic reptiles of the Mesozoic era in relation to their huge bodies. The balance can be restored if, in front of power machines, there is no longer a naked human brain, but a human brain *plus* information machines capable of playing the role of what, in nervous systems, performs automatic regulative functions. The relation:

$$\frac{\text{naked human brain}}{\text{weight of the organism + weight of power machines}}$$

is no better for humans than for microcephalic reptiles. But the relation:

$$\frac{\text{brain + automatic information machines}}{\text{power machines}}$$

tends to restore, on a higher plane, the good situation of the human who has not yet become a "vertebro-machine."

To be sure, Bergson's wish has not yet been fulfilled: this "supplement of brain" is not a "supplement of soul." Unfortunately, it is not yet the case that steam or atomic energy is self-controlled by servomechanisms and automatically guided toward wise and reasonable uses. But the supplement of the brain is already a highly significant good, and it is the primary condition for a supplement of soul. The human being who is freed from servile labor, whether cerebral or manual, at least has the opportunity to cultivate themselves and dominate their own destiny. Self-regulation, for a living being, is not wisdom, but it is the condition for and the beginning of wisdom.

The complete automation of industry is only the continuation of the very long evolution that has already automated the higher organisms. It leads to the replacement of physiological machinery by industrial machinery, to the elimination not only of the hand and the muscles but of the cerebral circuits that control them. In fairy tales, mysterious hands appear to be indispensable for carrying magic torches. In stylized torches, sculptured hands still carry the lamps. The doors of our apartments still have handles, and our electric lights are still controlled by switches that need to be maneuvered. But in a civilization where applied cybernetics will reign, lighting devices will illuminate themselves with the "information" of photoelectric cells. Doors will automatically open in front of us when we unknowingly cross an infrared beam. Human hands will no longer have to intervene in the apparatus of civilization, just as the I-consciousness will no longer have to intervene in the functioning of automated organic devices. The tool will still seem to extend the kinesthetic: we will still say of an ordinary machine or device that we have it well "in hand," its imperfect feedback*[15] still being completed by organic feedback. An automatic machine, however, functions without the intervention of humans. The soul and the hand have been removed from them, but in order to be liberated.

We know that in our motor cortex, the collection of all muscle control—as well as in our parietal cortex, the collection of kinesthetic centers of sensibility—constitute a kind of *Homunculus*. The proportions of this *Homunculus* are very different from those of *Homo*, since it represents the speaking and acting human rather than the living human. This *Homunculus* has a trunk and tiny legs, but an enormous tongue and hands. This is because, until the present, it was through the hand that humans touched the world of tools, or nonautomated machines, and through the tongue, the world of symbols representing concepts. However, cybernetic machines can in principle do without both the tongue and the hand. Machines that calculate and reason substitute the layout of their circuits for the arrangement of the signs of language, and allow us to realize the Leibnizian ideal: "Instead of disputing, let us calculate." Feedback machines, on the other hand, substitute their automatic circuits for manual manipulations, and the Baconian and Cartesian ideal: "Natural forces work by themselves like craftsmen, and replace the craftsmen."

THE THEORETICAL INTEREST

Cybernetics has no less theoretical interest. We have scientific understanding when we can create schematic models, when a technology can try to reproduce the phenomena to be known. Physiology and psychology have much to learn from the behavior of automata. The difficulties of production, with which technicians often struggle, attracts the attention of theorists and observers on the role and mode of action of the corresponding organs. A technology is often developed by modeling itself on certain physiological functions, broadly perceived; but the situation is reversed rather quickly, and it is the progress of technology that helps to improve our understanding of the physiological functions. Catalysts have helped us understand the role of diastases. The practice of photography has given us a better understanding of the mechanism of vision. The chemical study of buffer solutions has illuminated many aspects of organic metabolism. Ultrasound techniques have drawn attention to the way bats avoid obstacles through echolocation. There is no doubt that the practice of temporarily stopping the beating of the heart during surgery and replacing it with an automated pump will advance our knowledge of the physiological mechanisms of circulation. It was the circuits of mechanical feedback*[16] that attracted the attention of Rafael Lorente de No and helped him decipher the feedback circuits of nerve connections. It is the diverse oscillations of mechanical feedback that have helped us to disentangle the mechanism of shaking due to tabes dorsalis, on the one hand, and shaking that has a cerebellar origin, on the other. [It was after seeing the schematic of McCulloch's machine that von Bonin—rightly or wrongly—believed he could recognize the anatomy of the fourth layer of the visual cortex. The scanning* of systems analogous to television provides hope that we will be able to understand the observed link between vision and alpha waves. Even psychiatrists and theorists of neurosis and psychic conflicts probably have something to learn from the study of compound homeostats and their difficult search for equilibrium.

This list could be extended indefinitely. Cybernetics is already an indispensable propaedeutic for physiology and psychology. And psychological manuals inspired by cybernetics have already been written.]

THE POSTULATES OF CYBERNETICS BEFORE LOGIC

But it would be as dangerous to believe blindly in the models offered by cybernetics as it would be to disdain them. Mechanical models, or more generally schematic models, can be instructive on condition that they are used without dogmatism, and without presuming that everything in the physiology and psychology of nervous systems can be explained by models of this kind. They are instructive on condition that we expect as much illumination from their failures as from their successes, and provided that we do not claim in advance that every failure is merely apparent and temporary.

"Broad generalizations" are often dangerous, but under these circumstances, it is difficult not to agree with those who, more prudent than mechanistic cyberneticians, stand firm in the presumption that even the most perfect mechanical brain will always be, by definition, less perfect than the living brain, and that there will always be a gap between the two. On the one hand, the brain is self-fabricating; on the other hand, it is the brain that fabricates the automata that imitate it. The more the brain works marvels, the more its marvelous character becomes evident. Living human beings are, by definition, always a step ahead of their own products [*oeuvres*]; they can never be overtaken by them, since it is they who drive them forward.

Let us be clear. This "step ahead" that humans maintain is not quantitative; it is a difference of order, not of performance. Calculating machines can easily beat the best human calculators; a thyratron tube does a better job than the finest workers; an electric eye, like a camera, is far better in many ways than a living eye. Henri Dubreuil, in front of the sophisticated machinery of American industry, experienced a humanist satisfaction that was more touching than enlightened when he noted that the final testing of a manufactured object was still done by hand.[17] But Dubreuil's book is rather dated, and it is likely that today, in the same industrial establishment he visited, the hand has been replaced by an electrical sensor or inspector. There is no reason not to presume that machines will soon overtake organs in areas where the latter are for the moment still superior. But what is incontrovertible is that the living organism is first in relation to machines. Some machines are manufactured by other machines and are self-regulated, but it is the organism that initiates this chain and sustains it.

Better said, the organism itself is constituted by a set of organic machines, but there is something in the organism, or beyond the visible organism, that must still be first in relation to these organic machines, since it is that something that creates them. The word "organism" is profoundly equivocal: it designates, at one and the same time, the collection of organs but also the fabricating and utilizing unity of these organs. The fabrication of calculating and reasoning machines is secondary relative to the embryonic fabrication of a living brain. We could accept that the functioning of neural circuits and synaptic switches is of the same nature as the functioning of electrical circuits [and flip-flop* cells]. But this would simply prove, once again, what has long been known—namely, that there are machines in the organism—but not that the organized being is itself a machine. It would prove that the unobservable being that appears as the first human cell is able to build, without a machine, the organic machines that in turn are able to produce automated nonorganic machines, which themselves can control nonautomated machines. It would prove that what we call an organism is both something observable in space and an unobservable x that maintains the entire chain of internal and external automations.

Once the chain has been initiated, we can see that the circuits resemble each other: self-controlled electrical circuits are very similar to recurrent neural circuits; photographic devices resemble the eye; the lenses of glasses look like the lenses of eyes. But the first circuit must necessarily be of an entirely different order than the second. Nothing in the initial cell resembles a nervous system or a lens. Here, active information is absolute, even if one considers only space and time. The cell is original in every sense of the word.

This broad generalization is impossible to attack, and it has the serious drawback of constricting the mind without enlightening it. It can teach us nothing about the nature of the nonmechanical part of the living organism. It is as unassailable as the broad arguments against perpetual motion, which give us no reason to examine the proposed pseudo-solutions. In fact, the fundamentalist faith of cyberneticians in their models is not that different, psychologically speaking, from the faith of those who search for perpetual motion or try to square the circle. It is perhaps momentarily useful because of the enthusiasm it awakens, but it can quickly become harmful. In science, rational rejections are more helpful than excessive enthusiasms. The recognition of the impossibility of perpetual motion was a leap forward, as was the rejection of circle-squaring. Blind faith in cybernetic models may be useful to technicians since it can encourage them to try anything and everything. It can still be useful in some theoretical domains. For a long time to come, physiologists, in particular, will benefit from positing that everything is done by assembly [*montage*] and functioning, and from asking technicians to imitate physiological functions in their entirety. But blind faith is today harmful in other scientific domains, such as psychology and embryology. In philosophy, most notably, it distorts our vision and prevents us from seeing clearly paths that might lead to important advances in our conceptions of nature. It is time to seek out postulates and directions of thought that are better suited to psycho-organic reality since, as we will see, these directions are in perfect accord with the new directions of the physical sciences themselves. Despite its clearly "modern" spirit, cybernetics borrows almost exclusively from classical physics and not from microphysics. Its spatiotemporal postulates have already been abandoned by contemporary physicists. By making cybernetics less mechanistic, we will not be distancing it from the scientific point of view; on the contrary.

It might seem surprising—and unlikely—that the eminent mathematicians and physicists who founded cybernetics could have entertained such obviously false assumptions, as obvious as those that are the basis of the error of perpetual motion. We find ourselves wondering if it is not the critique that is superficial. The error, if there is one, is even more suspect, since it would have come from those who emphasized the close relation between entropy and information. Everyone knows that the discovery of the degradation of energy immediately raised problems of origin, problems that were previously foreign to physics. If entropy and information can be expressed in opposite but related formulas, what is valid for one is valid for the other, and problems of origin arise for both. It is curious that this consequence has not been noticed.

But it must be understood that the theorists as well as the technicians of cybernetics have little interest in anything that is not an immediate technical problem. Moreover, the critique of postulates will have little impact on them, because in this case, theirs is a semi-voluntary negligence rather than an error. At bottom, deterministic science before Planck and Heisenberg had always been highly indifferent to critiques that were purely philosophical or logical. There was also something contradictory in the postulate of determined functioning, whose linkages are everywhere but whose origin is nowhere, and which reduced all beings to nothing more than pure places of passage in an "infinite" causality. Determinism—it would be better to say "the theory of strict spatiotemporal functioning"—has simply succumbed to a palpable technical impossibility. Even today, in experimental psychology and physiology, experimenters continue to adopt the postulate of determinism in their studies of behavior. Even when, under the pressure of their experimental results, their descriptions increasingly recognize the inventive nature of behavior, a hidden determinism is still retained by stylistic convention. Cyberneticians reckon that they have no need to be more philosophical than the behaviorist psychologists.

Thermodynamics, even though it is deterministic in its postulates, has been forced to raise questions of origin, but it has done so for technical reasons, not for philosophical ones. Engineers have taken an interest in Carnot's principle and the spontaneous increase of entropy because the principle sets a vexing limit to the performance of machines and because fuel is scarce and expensive, not because it poses the problem of the origin of energy in the universe. Cybernetics, from this point of view, may be more negligent than thermodynamics, since information, the "supply" of information machines, does not seem as costly or threatened by increasing shortages as is coal or oil. The artistic director of a radio station may worry about finding good creators of information—good singers or good actors—and he knows their price may be rather high. Art critics can complain about the inanity of radio shows, and even if they are particularly pessimistic or demanding, they can always hope that nourishment will be found for this ravenous pit of invention, discovery, and talent that constitutes radio broadcasting. But the only worry of the technical director, or engineer, is to make sure there is a good modulation. The prospecting or "mining" of his country's artistic resources is not part of his job, and he is far less worried about the possible depletion of resources than is an engineer specializing in internal combustion engines.

Cybernetics is manufacturing automata and information machines that are becoming increasingly sophisticated. At the same time, it is finding more and more automations of the same kind in the physiology and psychology of living "informants." It is holding both ends of the chain, and pursuing its two orders of work, without worrying much about how they are connected. And without realizing that it is very improbable to admit, even implicitly as a postulate, that a day will come when automatic machines will not only ensure the dissemination of information but will fabricate it from scratch—in short, that a day will come when the technical director will serve as the artistic director, himself waiting to be replaced by an automatic director.

But this is only one half of the strange consequences contained in the postulates. We move from an automatic "informant" to one that is automatically "informed." The nervous system is supposed to be reducible to machines, in its receptive functions as much as its executive functions. Mechanical listening devices will have the advantage of not getting bored when broadcasts become monotonous and repetitive. At first, we might suppose that engineers, through a refinement of their desire to reproduce life exactly, will find a way to mechanically imitate boredom, just like everything else. As certain experiments by Pavlov suggested, boredom is probably a phenomenon of internal inhibition, engendered by the ineffectual repetition of the same stimulant, and this phenomenon can be perfectly imitated by a mechanism. It would suffice to use a counter-electromotive force that progressively increases until it begins to act at a certain threshold. Once this threshold is exceeded, automatic listening devices will go on strike and stop listening, or they will automatically send letters of protest to the automatic director of the broadcast. But we will soon discover that it is more expedient to eliminate such "mechanisms of boredom," and as a result we can stop renewing the stock of information coming from the transmitter. As we have seen, information machines use information as their supply, just as heat engines use coal, but the difference is that they do not necessarily degrade the information as such. If radio stations are immense consumers of information, it is for completely different reasons than those of the machines that are immense wasters of radiated energy. This is not because they are wasting the information being transmitted; it is because listeners quickly tire of what they have already seen or heard and need to have their "psychic nutrition" constantly renewed. Once the living listeners have been replaced by machines whose "organs of boredom" have been removed, then nothing will impede the realization of the perpetual motion of the third kind that is implicitly accepted by the postulates of cybernetics. Transmitters and receivers will always consume electricity: they will never contravene Carnot's principle and will never create a perpetual motion of the second kind. But they will make the same stock of information circulate in a circuit that could be closed by a "return" from informed automata to informing automata. The whole will resemble the automatic memories of calculating machines and their waves maintained in mercury tubes.

We thus see that, after all, perpetual motion of third kind is not the result of a mistake or blunder by cybernetics, but is implicated in the very definition of information as understood by cyberneticians. But we also see—or at least we hope—that the entirety of the mechanistic interpretation of the theory is reduced to an absurdity and vehemently requires revision. The circulation of waves in a closed circuit moving from machine to machine, with neither an origin in nor an exit from an individual consciousness, cannot be called information. Otherwise, we would have to say that ocean waves "inform" each other when they are born endlessly from one another. A kind of information that is remotely analogous to conscious information might perhaps be the basis of the phenomena of interaction in wave mechanics. But a cycle of coupled functionings is surely not information.

THE MYTHS OF CYBERNETICS

Journalistic and popular portrayals of cybernetics exhibit a number of mythical themes, almost like a psychoanalytic confession, though they are confused and contradictory.

In every era, automata—from the articulated gods of the Egyptians to ENIAC, and passing through the toys of Archytas, the clocks of famous characters in the Renaissance, and the mechanically animated figures of Jacques de Vaucanson and the Jaquet-Droz family—have always been surrounded by a kind of *aura*. This can be seen in the dark glow of the tales of E. T. A. Hoffmann, Edgar Allen Poe, and Villiers de l'Isle Adam. Automata seem to awaken in people both fear and pride.

Especially fear. When it is too perfect and able to function on its own, a machine created by a man becomes something foreign and hostile, since it can move toward or against him, and grasp his hand of flesh with a hand of iron. A worker can be caught by the blind functioning of an ordinary machine: a simple accident. But an engineer can be shot and killed by an automaton capable of finding and tracking a target but unable to recognize its own creator; this is more than an accident, it is a kind of tragedy, like Nemesis. The mechanical imitation of the situation of Frankenstein and his monster produces the same horror in us. This horror is intensified if the mechanical creature can calculate and reason with implacable rigor and without sentimental deviation. It would be intensified further if the human engineer allowed himself to be, not killed, but contaminated in his soul by the mechanical egoism of his masterpiece. The machine would then become a kind of damned soul of the man. Tyrants once had their astrologers, but we can imagine future tyrants consulting their "electronic brain," which would be capable of performing strategic calculations infallibly. An all-powerful head of state, calculating the risk of a war, fortunately knows that he may be wrong. But if a logistics machine invariably concludes that a war is advantageous, it will be harder for the tyrant to resist the temptation. Mechanical logic will become fatality.

Yet pride also. Despite the dangers of the operation, man has often dreamed of becoming a real demiurge, of creating beings that walk on their own and outrun him. Information machines fascinate him so much because they seem to mark a decisive progress toward the old dream of demiurgy. The utopia of the mechanical creature renews the theme of the creation of man by God, but by inverting it. The myth of a new original fall, possibly started by his own creature, though frightening, fascinates and flatters man, because it is he who would play the role of the God who is disobeyed. He is disappointed to be brought back to reason, when it is pointed out to him—what is nonetheless an obvious fact—that an electronic calculating machine is no more a superhuman brain than is a cogwheel, or that an airplane without a pilot or a self-guided missile has nothing more sinister in it than the thermostat for his apartment's heating system. He is disappointed when he is forced to realize that it is impossible to sever the ties that bind his creatures to his own life, when he is forced to see that his technologies fail to truly create, just as his science fails to truly understand.

Common sense tells man that if his own life did not have something mysterious and incalculable about it, he would not be a living being. But Thought, unique and in-itself Being, or Reason as a mystic faith, continue to postulate an absolute science and technology. As Henri Frankfort showed with regard to the Ionian physiologists, it is more difficult than is commonly imagined to escape from mythical thought.[18] Man and nature are linked together, and it is even more impossible to rationalize nature absolutely than it is to lift the earth with a lever without a fulcrum. There are as many myths in absolute rationalism as there are in religious imagery.

Speculatively and technically, the idea of an autonomous human reason, capable of judging the absolute origin of things and creating an independent life, is as much a mythical concept as is the idea of a creator god. It entails the belief in a kind of perpetual motion as the last word on the Whole of Being. From the seventeenth to the eighteenth century, when, as in ancient Greece, reason was gradually becoming secularized, a person who did mathematics, especially geometry, was thought to participate in divine reason. By the eighteenth century, he was simply participating in natural reason. Today, finally, mathematics is nothing more than simple conventional coherence. Calculators and information machines seem to represent the final, if extreme, term of this evolution. Aristotle attributed a rational soul to man, above all because he was capable of doing syllogisms and counting. It is true, as Bertrand Russell noted, that the Greek system of numeration was so bad that calculations were a genuine achievement. Today, when logical and arithmetic machines are much faster than the smartest human, it is hard to believe that these machines are immortal or are a part of divine spirit. But the "numinous," which is latent in absolute rationalism, is nonetheless subconsciously attributed to them.

One might also say that reasoning machines, or machines with finalized mechanical behavior, seem to confirm the philosophical fantasy of a pure human being as a "rational being" rather than a "rational animal." The word "animal" here disappears from the classical definition of man. In his science fiction novel *Last and First Men*, Olaf Stapledon imagines the future of humanity and related species for hundreds of millions of years. One of his themes is the Great Brains. Using the scientific processes of *in vitro* fertilization and directed embryonic development, the Man of the third kind strives to create giant brains, assisted by a host of auxiliary instruments and freed from any organic apparatus. Gone are the viscera that, in the present form of human beings, do little more than to distract the brain from its intellectual operations with all kinds of impulses and absurd emotions: "We must produce an organism which shall be no mere bundle of relics left over from its primitive ancestors and precariously ruled by a glimmer of intelligence. We must produce a man who is nothing but man."[19] If it is true that the most perfect characteristic of human nature is to have only adequate ideas, then is not the true man a thinking machine? It matters little if you get there by biological or surgical techniques, starting with a man of flesh, or by fabricating a brain directly and mechanically with electrical circuits. If it is true that "to be human" is essentially and exclusively "to be reasonable," then the thinking machine is more truly human than a pleasure-seeking or passionate person.

Chapter 1

The Main Types of Information Machines

In briefly reviewing the main types of information machines, we will consider them from two points of view: What is their value as models of psycho-biological actions? And what is their role in the use or possible extension of information? The two main types are, on the one hand, electronic calculation and reasoning machines, and on the other, feedback and self-regulation machines. Machines of the first type, which *process* information, belong in the field of informatics rather than of cybernetics.[1]

CALCULATING MACHINES

Your typical office calculating machines are based on cogwheels, which move according to the ratios determined by the decimal system. Analogue machines are mechanical models of the phenomenon that we want to study; they provide information such as length, rotation angle, and so on, which can be easily translated into measurements of the phenomenon. Machines for solving differential equations have an integrator consisting of a horizontal disc, a roller rotating on the surface of the disc, and a screw that moves the center of the disc relative to the wheel. The rotation of the wheel depends on both its distance from the center of the disc and the rotation of the disc. If, for example, the displacement of the center of the disc measures the speed of a vehicle, and if the rotation of the disc measures time, the wheel will measure the distance covered.

Since there is obviously nothing in the brain like cogwheels, screws, or discs, no one can be tempted to find a key to understanding how the brain works in these kinds of machines. Moreover, calculating machines have almost never interested physiologists or psychologists as much as have automata proper. Already in the seventeenth century, Descartes, the godfather of cybernetics, derived the reflex model and all the essentials of his physiological conceptions from the simplest of these hydraulically controlled automata. But it was not until the twentieth century that calculating machines and analogical machines became interesting to philosophers after Pascal and Leibniz, other than for their logistical aspect.

Electronic calculation machines (ENIAC, BINAC, EDVAC, Mark 1, etc.) are designed on a completely different principle, that of electrical circuits that are controlled, that is, kept open or closed, by electronic valves similar to radio lamps. These machines operate not by mechanical movements but by opening or closing circuits. The *Current–No Current* alternative represents basic information, the *Yes–No* of an informative decision. With basic *Yes–No*s multiplied, one can express or learn anything, give or receive any information. With *Yes–No* questions in a board game, one can lead an initially clueless player to find a hidden object, a name, a number, or any arbitrary concept. A photographic or televised image is only a set of well-coordinated, luminous *Yes–No*s. The luminous intensity at each point can in fact be reduced to *Yes–No* responses accumulated over time.

This principle can be used directly for numerical calculations, without the need for cogwheels and gears. This is particularly simple if we replace the decimal system with the binary system, on which Leibniz, haunted by the idea of a universal calculus, had already meditated. In the binary system, the dyad replaces the decade, and all numbers can be written with only two signs, 1 and 0, Yes or No. Two will be written "One dyad, zero units," or 10. Three: "One dyad, one unit," or 11. Four: "A dyad of dyads," or 100. Eight will be written 1,000; Sixteen, 10,000. With four information units, we can write the first fifteen numbers. In binary notation, the numbers are on average three times longer than in decimal numbering, but for machines that make thousands of additions to the second digit, this is of little importance. And the addition and multiplication tables have an admirable simplicity:

$$1 + 0 = 1 \qquad 1 \times 0 = 0$$

$$1 + 1 = 10 \qquad 1 \times 1 = 1$$

For addition, and for the other operations which follow, the electronic machines use a so-called flip-flop* assembly, which uses two switching valves: closing the circuit of one opens that of the other.

The flip-flop cells, each having two rocker arm valves, are stacked, and each level, by means of the rocker arm, allows one pulse out of every two to pass. This consequently corresponds to the writing system of binary numeration, in which the number of units changes after each successive number; the number of dyads after two numbers, the number of dyads of dyads after four numbers, and so forth. The other operations are easily derived from addition. It is also quite simple to transpose the results into decimal numbers.

REASONING MACHINES

[Similar arrangements can be used for logical calculations with a True–False alternative. However, there is no way that a machine can answer "True" or "False"—even in machine language, and assuming that translation is possible—if one gives it an isolated proposition such as "The sun revolves around the earth" or "Pelagianism is a heresy." The logical truth, unlike the factual truth, is always hypothetical. The logical "True" and "False" are always coordinated by "If's." A machine can deduce; it can calculate the logical truth, that is, it can carry the truth or falsehood from one proposition to another and operate on V and F as it operates on numbers. To do this, it is sufficient to make set-ups corresponding to the logical articulations of the propositions: *either, and, no, or otherwise, if . .. then, if . .. and only if, some, everything,* and so on.]The English mathematician George Boole, around the middle of the nineteenth century, showed that we can reduce logical reasoning to an algebraic calculation of logical reasoning.[2] [By representing the value of the truth of a proposition by 1 and its falsehood by 0, we can make all kinds of operations on the True and the False.[3] The main logical expressions are represented by formulas. If we let P, Q equal propositions, and p, q, their truth value, we will have the following:[4]

PROPOSITIONS	TRUTH VALUE
Not-P.	$1 - p$
P and Q	pq
P or Q.	$p + q - pq$
If P, then Q.	$1 - p + pq$
P, if and only if Q.	$1 - p - q + 2pq$
P, or otherwise Q.	$p + q - 2pq$

For example, take the formula $p + q - pq$. It represents the conjunction "or." Indeed $p + q - pq$ must have a truth value only if P or Q is true. This is easy to verify:

	pq	$p+q-pq$
P and Q both true:	11	$1+1-1=1$
Q true	01	$0+1-0=1$
P true	10	$1+0-0=1$
P and Q both false:	00	$0+0-0=0$]

Claude E. Shannon, in 1938, showed that Boole's algebra could be translated into combinations of open or closed circuits.[5] [To represent the conjunction "or" it is sufficient to have two switches connected in parallel: The current flows (= 1, yes, true), when either one or the other, or both switches, are closed. This is the equivalent of "P or Q." Two switches in series will be equivalent to "and," and so forth.]

Using this correspondence of principle, Theodore Kalin and William Burckhardt built a logical calculation machine in 1941[, which was actually used in the calculations of an insurance company.]

There is obviously no way that calculating machines, whether arithmetic or logical, can increase information. [If the operator is mistaken in assigning a truth value to P or Q, the machine will deduce the false as well as the true, and will not restore the truth.] But undoubtedly, such machines may appear similar to the brain. There are indeed circuits and valves or switches in the brain. The brain may well be only a set of switches and circuits, in which a current, which is not electric, but electrochemical, always moves in the same direction. The nervous switches, or synapses, between neurons function like electronic valves, according to an all-or-nothing law. Of course, it is highly doubtful that the brain circuits, while working in a given operation, are precisely set up like the electrical circuits of a machine when it is performing the same operation.

When we add, we do not use the binary system, and our neurons do not stack up like "flip-flop's."

When we think with "and," "or," "if . . . then," "but . . . ," it is very likely that we do not simply put nervous circuits in series or in parallel. We have, as William James says, the "feeling of but" or the "feeling of because." Nevertheless, the fact remains that the two operations, physical or physiological, may be similar, if the atmosphere of meaning that envelops the cerebral operations is as insignificant as the emotional atmosphere that often envelops the reflexes. If I brake suddenly in front of an obstacle, my fear, which often arrives late, has nothing to do with the assembly that allowed the prompt and appropriate response.

A bus driver, at a stop in front of a crossroads, hears the conductor's bell that allows him to depart. But the traffic light at the crossroads is red. He only restarts the vehicle when the light turns green. There must be some kind of serialization in his cortex, equivalent to the "and" of logical machines. Similarly, the driver in the moving bus prepares to stop at the next station either if he has heard the bell ringing from a passenger requesting the stop, or, with or without the bell, if he sees passengers on the sidewalk. It is the equivalent of the "or" of logical machines. The experienced driver has no need for an *atmosphere of meaning*[6] around these two operations, and an automatic assembly could very effectively replace him.

Calculating and reasoning machines also imitate psychological operations auxiliary to logical or arithmetical calculation. When a man does a multiplication, he applies learned rules, he consults his memory of the Pythagorean table, and he also entrusts to his memory the partial results obtained. On the other hand, he must be able to read or listen to the data of the problem and express the results obtained. Perception, memory, the consultation of memory, the application of rules, and expression: all this can be given to machines. They perceive the problems to be solved and their data are transcribed on perforated tapes; they have memories (magnetic tape, continuous wave tubes) that they know how to consult at the right time. Not only do they obey instructions, but they can even choose which part of the rules they apply, thanks to selectors that automatically decide for a series of auxiliary operations, according to the needs of a more general operation, directly controlled by the machine. Finally, they give the printed and verified results.

We could easily complicate the presentation of a calculating or reasoning machine, by making it truly read instructions by means of a photoelectric eye and by making it speak the results by means of a *Voder*.*[7] The "mentality" or "thought" of a machine would still be a pure metaphor. The machine can only function; it can never itself determine the totality of the rules it applies, but only a part that is strictly predetermined by its set of assemblies and not really chosen.

Edmund C. Berkeley provides the schematics for the assembly of a basic calculating machine that he calls "Simon," and even offers to send additional instructions to amateurs.[8] This machine uses only a few dozen valves and a few dozen meters of wire. It is able to teach us great truths such as $2 + 1 = 3$ and $2 - 1 = 1$. It is able to notice that the first of these results is the largest, or to choose such an operation, depending on whether the first of the two results is larger or smaller than the second. To get a reasonable idea of the ENIAC or EDVAC, it is well to remember that they are only larger versions of Simon.

INDUCING MACHINES

Inducing machines are still largely at the early stage of planning. This is unfortunate, because induction, unlike deduction, seems to be an enriching operation as far as information is concerned, and therefore very closely concerns the problem of its origin. As cyberneticians work relentlessly on the problem of learning,* which, as we will see, involves a certain statistical induction, we can imagine that machines that can draw general laws from statistical data will soon appear. Already in 1925, the American behaviorist C. L. Hull invented a machine for making correlation calculations.[9] This is a matter of calculation, and therefore nothing extraordinary for a machine to be able to do. But, on the other hand, a calculation of correlation is very close to an induction. Do tall men also have a wide arm span? To answer this question, you need to take a tape measure and measure the height and arm span of a large number of subjects. By comparing the scores of the subjects, and applying the well-known formulas, we find that the correlation is +0.82. The machine can do the calculation when given the results, which is comparable to the extraction of a general law from particular data, while also accompanying the law with a quantitative degree of probability.

[Provided it is given sufficient data, the same kind of machine could theoretically answer the question "Do animals without bile live longer than others?," or "Are blond men with blue eyes more enterprising than others?," or "Are short and fat men more prone to cyclothymia?"]

Another operation, very similar to induction, is interpolation, or extrapolation, and several logicians have reduced induction to interpolation. A machine can easily interpolate and extrapolate; interpolation between two numbers in a table is one of the common operations of electronic calculation machines. The simplest possible example of an interpolation "machine" is provided by an ordinary ruler, with the help of which a straight line is drawn or extended from two given points. Other examples that are almost as simple are a compass, a reduction compass, a curvigraph, and a drafting pencil. A machine using these simple instruments may be able to extend a series of numbers obeying a rule, or to fill a gap in the series:

$$5 \quad 7 \quad 9 \quad 11 \quad 13 \quad 15 \quad (\ldots)$$
$$100 \quad 90 \quad 81 \quad 73 \quad 66 \quad (\ldots) \quad 55$$

It can automatically indicate the rule of the series. It can also correct a "false" number in a series of numbers. The "false" number in the series 4, 11, 13, 19, 20, 19, 16 or 0, 0, 5, 9, 12, 14, 15, 14, is not obvious, but a machine that could translate them into graphs and could detect anomalies in the curves by pressure on a membrane could easily detect it. However, all these operations are considered to be intelligent operations, requiring a certain amount of inventive effort, as they are classic tests of intellectual level. Machines are capable of a host of "eductions of relation" from given correlates, or "eductions of correlates" according to a given relation, to use Spearman's vocabulary. These constitute actions that Spearman considers to be typical of intelligence and as "noegenetic," that is, as involving the creation of knowledge. Spearman gives as an example of an "education of correlates"[10] the task of finishing a figure according to a model of different proportions (figure 1.1).

It is possible to conceive an amplifying machine that would not only reproduce the *ab* part by enlarging it, but which, conversely, when the path $a'b'$ is imposed on it, would recognize the degree of amplification by a kind of automatic learning, and would continue the curve according to this degree until c' is reached. By combining the opticians' devices to measure the number of diopters of a spectacle lens with a lens-cutting device, we would have an "intelligent" machine, at least if we take Spearman's definition literally. An interpolating machine could have aided Kepler, and even, in principle, could have replaced him in the task of deriving the laws governing planetary trajectories from Tycho Brahe's observations. Today, in fact, radars with automatic computers detect the trajectory of large caliber mortar shells and determine their point of origin by extrapolation.[11]

Completing or restoring information is an operation very similar to—almost indistinguishable from—the creation of information, or invention. Any invention, even the most seemingly spontaneous, can be considered as a recovery of information. When psychologists have tried to discover the origin of invention, they have in fact used tests to restore incomplete or confused information: sentences or words to be completed, series of drawings to be put in order, and so forth. In science or technology as well, invention is almost always a completion or restoration of information. Cyberneticians may have some hope of making inventing machines. Moreover, it is not improbable that the inventor's brain often works like an interpolating machine, that

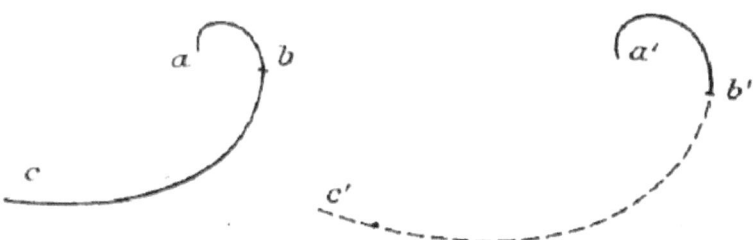

Figure 1.1.

is, through harmonious completion, the rounding of data, and the dynamic search for simplicity, symmetry, and equilibrium.

On this point, as on many others, cybernetics can have recourse to that more subtle behaviorism that is Gestalt Theory. We know that for this theory intelligent invention represents neither a transcendent action nor a process of pure trial and error without direction, but a spontaneous reorganization of the perceptual field, isomorphic to the cortical dynamic field. This reorganization is carried out according to the principle of least action or of least tension.

It is very interesting to realize that, if taken to the logical conclusion of their fundamental hypotheses, cybernetics combined with Gestalt Theory again leads to perpetual motion of the third kind. We know that in macroscopic physics, the principle of least action has the closest relation to the principle of evolution toward maximum entropy. To bring invention, creation, or the restoration of information back to the principle of least action, is in fact to bring information back to what, according to cybernetics itself, is precisely its opposite, entropy. It is the increase in entropy, the march toward symmetry, rounding, and equilibrium, that would ultimately explain the increase in information, whereas information is entropy with the opposite sign. The contradiction is obvious.

Moreover, experience shows that in the history of ideas, the "economy of thought" (as Ernst Mach used to say, anticipating Gestalt Theory), the search for a harmonious, well-rounded theory, leads much more often to an unfortunate stop than to progress. Information and theories that are too "beautiful," too simple, are almost always false. As for invention itself, for example in the order of technology, it is all too paradoxical to interpret the ever-increasing complexity of industrial devices as a march toward "good forms." For a machine or an organism to improve almost always means for it to become more complex. What can be misleading, however, is that after inventing a new, more sophisticated and complex device, the inventor—and this is part of the engineer's intellectual reflexes—seeks to eliminate unnecessary complexities. This can be done by making a single mechanism useful for two purposes, if they are needed. In cars today, the horn control is mounted on the same lever as the lighting control, just as in vertebrate organisms, the genitourinary system is mixed. Successful machines, like successful organisms, have a simplified, rounded, smooth, external appearance, without add-ons. Modern airplanes, which are so complex, seem simpler to the eye than the old biplanes of Farman's time. A dolphin seems simpler than a sea cucumber, which is in reality more rudimentary. Economy in invention, far from being its very essence, is only a last and late improvement.

We can therefore suspect that if an interpolating or inducing machine seems to imitate certain intelligent acts according to Spearman's definition, it is not because the machine is intelligent, it is because the definition is poor. Spearman is the opposite of a mechanist in psychology, and he understands that "noegenesis" transcends the plane of mechanical functioning. But he is wrong not to distinguish sufficiently, in his definition, between what is an understanding of the *meaning* of a relation, and what is imitable and feasible as a rule inscribed in a machine. The test of completing words in a sentence only seems to resemble number completion or the extension of a curve. The machine for detecting errors (or pseudo-errors) in a digital series is in reality only a machine for detecting singular points, which may or may not have the character of an error, depending on the meaning given to it by a conscious observer. On the contrary, it can be an interesting "residue," in the sense that logicians of induction understand the word. Moreover, inducing machines can only ever serve as auxiliaries for consciousness; they cannot replace it. When astronomers use a stereoscope, with two photographs taken some time apart, to detect a small planet that has moved in the stellar field and which consequently seems to come out of the photographs, they are indeed using a semi-mechanical effect, but with a purpose very different than that of the search for an error. A machine that would correct all anomalies would not be a very good machine for discovering or inventing. If, instead of Kepler's mind, and his trust in Tycho Brahe the observer, a trust stronger than his mystical faith in Pythagoras, we had had a machine to interpolate and regularize the *Gestalten*, this machine would have had a high risk of converting Tycho's data into points on a circle, instead of revealing the elliptical form of the trajectories. In order for us to speak of true induction or invention, it is necessary that the mind interprets according to a meaning. We know that one of the serious inconveniences of automatic machines with a "probe head," for anti-aircraft battle, is that they are always susceptible to being mistaken about the identity of their targets and of confusing friend and enemy. Appropriate signals can guide them, but accidents are frequent.

The radars that defended Antwerp against low-flying bombs apparently had the tendency to mistake their targets and adjust their aim toward the surrounding church steeples.[12]

SELF-REGULATING MACHINES

In self-regulating machines, the information doesn't simply pass from the input to the output of the machine with one-way transformations, as in calculating machines; it is used in a recurrent circuit that goes from the output back to the input, in feedback, to control the very operation of the machine.

Let us first consider the intermediate case of a so-called subservient mechanism. For example, the rudder control wheel of a large ship does not act directly on the rudder; its movement triggers the operation of servomotors that rotate the rudder until it reaches the desired position. The servomotors then stop automatically. The information given by the helmsman is used by the machine, which is responsible for operating according to the information. The same is true for the control of an elevator. If I press the "Fifth Floor" button, I give a certain "information" to the device, which knows how to use it.

Now suppose that the helmsman of the boat is replaced by a gyroscopic "heading indicator." The helmsman, or the captain, will only give information in the form of a general "ideal"; he will not have to read the deviation relative to the fixed heading at each moment. The gyroscope will give all the intermediate and technical information needed to reach the destination. If the boat deviates from its course, it is the gyroscope—or more precisely, a device for "reading" the difference between the fixed heading and the actual direction of the boat—that will be informed, and that will then inform the servomotors.

The self-regulating machine will apparently not only operate, it will also perform actions that are finalized, in the etymological sense of the word. Everything will happen as if an ideal end controlled and corrected its operation, by self-regulation, until it reaches that end.

[We see that automata of this type do not imitate pure thought—that is, the purely combinatorial aspects of thought. Rather, they imitate what we could call "interested consciousness," governed by an instinct or a value, which seeks a reduction of tension between its current state and its ideal state. From this point of view, they do not go at all in the direction of the fantasy of being rational in the pure sense. They imitate instinct or tendency, not reasoning. Through its instincts, an animal is sensitive to certain objects or situations: sugar, light, heat, humidity; it seeks them or avoids them. A man has instincts and, moreover, he is sensitive to certain values that give him affective interests and that largely guide his actions, while leaving the choice of means, which depends on intermediate information, to his control. It often happens that the various instincts or interests do not match. The animal or man is then obliged to choose or adopt an average behavior that harmonizes the various interests. Instincts and interests are not pure thrusts *a tergo*; they "compel" the organism to go in a certain general direction, but the intermediate behavior is flexible and can be adapted to circumstances and obstacles. Moreover, the animal or man is capable not only of improvising intermediate behaviors according to his purpose but of learning better ways to reach that purpose and of transferring the value of sign or of means from one intermediate object to another. Self-regulation, self-awareness, and the perception of interesting objects or situations; the balance and harmonization of interests; and even learning and transfer: all of these things, according to the cyberneticians, are common to both automatic machines and organisms.[13]]

FEEDBACK AND SELF-REGULATION

It was James Clerk Maxwell who, in 1868, made the first theoretical study of an industrial feedback system. He analyzed the operation of the ball regulator, which ensures a constant speed for the steam engine, despite the differences in load. It is essentially a pendulum, attached to a steam inlet. If the machine runs too fast, the centrifugal force lifts the balls that, through intermediate connections, decrease the intake. If the machine, loaded or braked, slows down, the centrifugal force decreases and the balls, falling by their own weight or under the action of a spring, increase the intake. The regulator, "informed" by the speed of rotation, transmits the "information" to the machine, which decreases or increases its speed accordingly. The new speed "informs" the controller again, and so on. The regulator acts in the opposite direction to that of the machine's operation; it corrects its deviations and tends to minimize its oscillations. This kind of feedback is called negative feedback.[14] A thermostat is of the same type: it regulates the temperature of the boiler. The same goes for the automatic piloting devices of aircraft. If the aircraft rolls, this informs the vertical gyroscope's artificial horizon, which operates the valve for controlling the fins, and straightens the plane. If this adjustment is excessive, a second action begins, which minimizes the oscillation. The study of the various types of oscillations in negative feedback poses delicate mathematical and physical problems, which we do not need to enter into here.

In positive feedback, the set-up is inverted [and the device uses the information to avoid, rather than search for, a certain level of functioning or an object that it is able to "perceive." The use of the words "positive" or "negative" seems rather unfortunate, but is explained because, in positive feedback,] the recurrent energy acts in the same way as the operation of the machine which races or returns to zero (runaway*)[15]. In both cases, the operation is not the result of a single thrust *a tergo*: a cause A (the primary operation of the machine) produces an effect B, and another effect b, proportional to B. The secondary effect b is related to A on which it produces a recurrent action. More precisely, b is automatically compared by the machine to an "ideal" bi, inscribed in the device by the engineer or user, and it is the difference $bi - b$ which produces a recurrent action on A. As a result, A becomes A' and produces the effect $B' + b'$. Thus, the automaton is mechanically finalized: by means of trial and error, it seems to seek an ideal state.

In any feedback, the recurring $bi - b \to A$ current can be called the "informant" of A. For example, while conducting an artillery shot, an officer at the forward observation post telephones to the battery commander: "Too long" or "Too short." A radar, registering the image of the goal and the image of the shot each time, can correct the pointing device just as well, by $bi - b \to A$ feedback. The recurrent flow is automatic information, which plays exactly the same role as the information given by a conscious observer.

If we consider each link in the feedback cycle separately, we find pure causality, that is, thrusts *a tergo*. But, as it is a cycle, what is behind each link is also in front of it: the guiding ideal combines with the thrusts *a tergo*, and a de facto finality seems to emerge from the mechanical causality. However, this cyclical nature, on its own, is not enough to characterize feedback. Information not contained in the machine's structure must be introduced into the cycle. Indeed, it can be said that any operation of any machine is cyclical, since the machine can always return to its starting point. We speak of the cycle of an internal combustion engine or a steam engine, even without a regulator. But, in an ordinary machine, the different parts are simply arranged to order each other in a specific way and at a specific time. If, for example, a cam mounted on the axle of a wheel lifts a lever arm each time it turns, it cannot be said that the cam "informs" the lever, without stretching the meaning of the word to the point of absurdity. The cam simply pushes the lever, in a conventional way, and in accord with the structure of the machine. In the combustion engine, the spark that produces the explosion that drives the engine, that controls the magneto, is produced by the magneto, and so on indefinitely. In a feedback machine, the recurrent flow of information also acts through *a tergo* thrust, but this time the information is not contained in the machine's structure. The ball regulator undergoes a thrust, and the intake register also undergoes the thrust of the regulator. What is new here is that the thrust information given by the regulator is not inscribed in advance in the machine's structure. It takes into account the various internal or external incidents that can change the engine's running speed.

Therefore, we can now define a feedback machine as *a cyclical functioning with a regulating loop through which a current of information flows, automatically compared to an "ideal."* In a machine without feedback, the conscious supervisor plays the role of the regulating loop; the supervisor observes the standardized functioning, compares it with the ideal to which he has the task of keeping, and intervenes to bring the standardized functioning closer to this ideal. In order to regulate, feedback performs these three functions: (1) it observes, (2) it compares, and (3) it intervenes.

One must not confuse the progress made in the primary functioning cycle—when an improvement in the structural arrangement of the machine makes it possible to do without the muscles of a worker—with the progress made with the regulating loop. This loop, which was previously represented by a supervisor and is now performed by feedback, makes it possible to do without the brain of a supervisor.

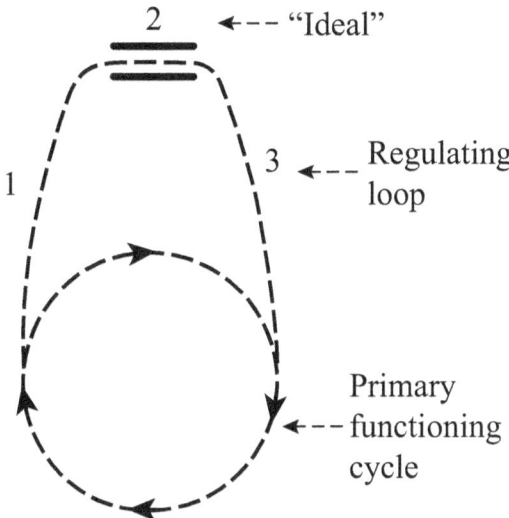

Figure 1.2.

ORGANIC FEEDBACK

The living organism is full of such self-regulations and physiologists very quickly adopted the notion of feedback that industrial technology offered them. This was generally in the hope of giving a mechanical explanation of the actual finalization of organisms. For a long time, moreover, these feedback systems had been defined under other names. Herman J. Jordan used the term "amboception" to emphasize the difference between an isolated causal action and the connected "multi-causality" or "causal framework." An apple falls from an apple tree and kills a mouse that was there: a causal action. A stone is supported by three pieces of wood; between these three pieces of wood is a piece of bait: a mouse gnaws at the bait, the stone falls and kills the mouse. In the trap, there is a "causal framework" and "amboception"; the bait is held on one side by the pieces of wood and on the other by the mouse.[16] In the organism, the secretin of the duodenum, for example, acts as an "amboceptor" between the acidity of gastric juice that, passing through the intestine, would prevent intestinal digestion, and pancreatic juice, which neutralizes gastric acidity.

We know that the equilibria in the body are, in general, not static but homeostatic; they remain stable despite variable external conditions. If the organic balance were static, and simply measured the resulting average of internal and external forces, life would be very precarious. In fact, organic equilibrium is generally not a result. It fully maintains a certain optimum value. Our blood maintains exactly the same acidity, even if we drink vinegar. It doesn't become more fluid, even if we have just absorbed a liter of water. Homeotherms keep the same constant temperature, at the equator or the polar regions. All homeostatic processes represent feedback, generally operated by slow-acting systems, such as the sympathetic and parasympathetic nervous systems, glands and smooth muscles, chemical messengers, and buffer solutions.[17]

Voluntary acts, on the other hand, represent rapid feedback, involving striated muscles and circuits of the central nervous system. Gestalt Theory, of which we have emphasized the similarity to cybernetics (and will do so again more than once), has described this feedback under the name of "circular processes." I am in the dark, and a light, appearing on the periphery of my visual field, makes me turn my head and eyes, then incites me to walk toward it. The functioning of my oculogyric and cephalogic centers, and then my locomotor muscles, is dynamically controlled by the result already obtained, and by the residual tension between this result and the goal. The Gestaltists reject the explanations according to preformed nervous arcs, which they think would imply an improbably large number of prepared assemblies [*montages*]. Cortical centers are not sets of isolated and juxtaposed pathways; there are "free" dynamic interactions between all these pathways, similar to those of a magnetic field between the spirals of an induction coil. [Cortical centers act in their unity, like fields of forces that reequilibrate on their own. The final command to the effectors depends each time on this reequilibration and can be improvised in detail. As long as I am not near the light, my field of behavior is still out of equilibrium, and any movement of approach by any mode of locomotion that improves equilibrium will be favored. The "circular process," while very similar to feedback, does not have a specialized conductor. The information is not brought back to the entry of the system by specific recurrent fibers; it is entirely one with the dynamism of the field of behavior.]

Cybernetics, in its orthodox form, reverts to the more classical theory of nervous circulation as subservient to conductors. If I want to take a full glass on the table, the pyramidal cells of the motor zone must be able to send impulses to the muscles of my arm. But it is not enough that I am not paralyzed and that I have good effectors. The central nervous system must be informed at each instant of the effect already obtained by the first impulse, and this information must be combined with that from the other sensory organs, so that subsequent impulses correctly complete the action. If the recurrent circuits of information were damaged, my movements would be clumsy and conventional, like walking in cases of tabes dorsalis, where there is no paralysis, but recurrent information is lacking. [The recent discoveries of physiologists, especially those of Lorente de No, seem to confirm the thesis of orthodox cyberneticians against the Gestalt thesis: there are in fact recurrent fibers of information coupled with fibers of effection.

The feedback arrangement of the conductors allows us to avoid the difficulty, emphasized by the Gestaltists, for any theory appealing to "assemblies" and "rails." Theoretically, a single reflex arc—provided it is controlled by a regular impulse—can perform different actions, depending on the information received. Grey Walter's artificial animals move toward the light wherever it appears, without having the multitude of conductors that Koffka considers necessary in the hypothesis of nervous paths serving as rails. Nor does the D. C. A. cannon with automatic aim have any such "rails."] A motorist can drive a car correctly, almost without error, with steering that is more augmented than what he is used to. This is because it is guided by the result achieved and not by a one-way motor habit. In fact, an automobile driver is unable to say, even very approximately, by what angle he should turn the steering wheel of his familiar car to take an intersecting road on his right. This is a striking indication of the completely improvised nature of the movements of effection. An animal that is mutilated after having been taught to leave a labyrinth, so as to prohibit some of the learned movements, improvises compensating movements and arrives at the result in another way, provided that some of the recurrent circuits are intact.

However, there is not much difference between the circular process of Gestaltists and the feedback conductors of cyberneticians. Feedback is, after all, only a circular process "on rails," a channeled dynamism. The Sperry Gyro[18] of the automatic pilot perfectly meets the requirements of the circular process, as well as those of feedback. The two schemas have this essential feature in common: they explain finalist regulations, "molar" behaviors, as predictable according to their aim, independently of the detail of "molecular" functions, without appealing to anything more than efficient causality. One can say of both the feedback system and the circular process system that they can be automatic and finalized at the same time.

Both cybernetics and Gestalt Theory stand in absolute opposition to the interpretations of introspective psychology. The latter see, in the de facto purpose of behaviors, the manifestation of a transcendent principle, which is irreducible to explanations or assemblies according to physical, mechanical, or dynamic laws. Clark L. Hull, among others, has stated with particular clarity, for the use of psychologists, what could be called *The Principle of the Possible Robot*: "Nothing should be assumed to happen in an organism that cannot also happen in a fully automatic robot."[19]

INTERESTED SENSITIVITY AND PERCEPTION

The "interested" sensitivity of a thermostat concerns the temperature: its organ of "perception" is a simple thermometer. The Watt regulator is speed-sensitive and "feels" the centrifugal force. We are not very impressed by these metaphors, because in these devices there is no staging [*mise en scene*] that can remind us of the way a living being stages the perception of an external object. If the information is provided to the automaton by a photoelectric cell, and if the automaton is equipped by its manufacturer with phototropism, the effect is much more surprising. Grey Walter's artificial tortoise, a three-wheeled device powered by electric motors and batteries, is topped by a photoelectric cell that controls the motors by relay to direct the automaton into the light. Moreover, the tortoise is endowed with a "stimulus" ["*tactisme*"]: when the shell hits an obstacle, a circuit closes, which neutralizes the phototropism, and releases the steered wheel for a moment: the device then seems to grope, scalloping around the obstacle, before resuming its path toward the light. The photoelectric cell is also sensitive to the brightness of light. If the light source is too bright, it acts as an obstacle and the tortoise avoids it. But when its battery has been sufficiently drained, the bright light no longer activates the "obstacle effect," and the tortoise moves toward it. If a bright light shines on a battery charging device, the tortoise seems to go after the electrical food, until the photoelectric cell can again act to move it away from the bright light. The automaton thus seems to imitate both the internal and external sensitivity of living beings, and what Edward C. Tolman calls psycho-organic "variable demands," such as hunger.[20]

Despite their quite spectacular character, Grey Walter's "animals" are rather crude devices. Their reactions have an "all or nothing" character, despite the twofold sensitivity of the cell. The perceived object, light or obstacle is only a trigger; it has none of the characteristics of a signal or sign acting by its form—that is, literally, providing information. Most industrial servomechanisms that use the sensitive "perception" of photoelectric cells are of the same order. They can be used to protect an installation, the device triggering an alert apparatus when someone approaches a capacitance plate or breaks an infrared beam. Or—and this is already something of a higher order—to automatically control delicate or dangerous practical work, and to guide the work of a machine according to a template whose form is examined step by step by the cell.

Even in this last remarkable case, where the machine seems capable of seeing a form, it is clear that it does not perceive, that it does not "recognize" the form as a form, since it limits itself to following the contour step by step. Cyberneticians very quickly grasped the importance of the problem of the perception and recognition of forms as universals,* that is, as overall schemas that psycho-physiologically preserve their identity, despite the different transformations of their geometric appearance, in the narrow sense of the word.[21] We recognize a familiar object, whether we see it from afar or up close, from the front or from the side. We recognize a circle as a circle, and a plate as circular, whether it is near or far, and therefore whether it projects a large or small circle on the retina. We recognize it whether it is seen from the front, side, or at a three-quarter angle; that is, whether it projects a circle, an ellipse, or an elongated trapezium on the retina. According to the assumptions of cybernetics, this performance must be done by nervous assemblies, and it must be imitable by mechanical assemblies. The reading machine of Pitts and McCulloch, to which we have already alluded, carries out, using appropriate relays, an independent scanning*[22] of the dimension of the letters to be perceived. But it is much more difficult to imagine a machine that would be capable of recognizing a plate, whether it is seen from the front, the side, or on a three-quarter angle. In short, this would be a machine that would work on similarity and that would be able—to use the words by which Kant defines the role of the schema, and that W. Russell Brain takes up again—to "subsume an object under a concept." Grey Walter's tortoise heads for its "hutch" when its battery is flat, because this hutch is mounted with a bright light, which triggers the power unit. But it does not recognize the form of its hutch from any perspective.

Norbert Wiener has tried to improve the way the problem is posed, by noting the following:

1. That the motor apparatus of the eye and the adjoining brain centers are set up, not only to bring onto the fovea an object whose image is first peripheral, but—if it is an object of which we are more familiar with one orientation rather than another—to put its image in this privileged orientation. It is unpleasant and difficult to read a page that has been placed wrongly, and we straighten it up as quickly and instinctively as possible.
2. That we recognize an object above all by its contours, where there is a contrast between two regions of different color or brightness.
3. Finally, and above all, that the different transformations of a shape according to perspective constitute a "group" in the mathematical sense of the word, a group itself comprising several subgroups: homogeneous enlargement, affine group, rotation along various axes, translation, and so forth. As a result, just as the Pitts machine, or the "reader" of television, explores and scans an ordinary two-dimensional surface, machines can be designed to explore the various subgroups of transformation involved in the various perspectives of a given object. Such

mechanisms may exist in the brain, combined with the center of foveal fixation. When operating, these group scanning*[23] mechanisms must necessarily fall on one of the possible transformations of a standard model, stored in our (physiological) memory, or at least fall within a certain margin of tolerance in order to resonate with the "model." This will then allow us to recognize an object as similar, as universal, and to act accordingly. Grey Walter's automaton, equipped with such relay-mounted group scannings, could then recognize its hutch from any point of view.

Let's even suppose that one of these group scannings is damaged. The automaton will then appear to suffer from a specific "agnosia" quite similar to the various agnosies that are at the root, in particular, of the different forms of sensory aphasia. We know how difficult it is to classify aphasics using psychological terms, or properly linguistic classifications. The varieties of aphasia described by Henry Head (verbal, nominal, syntactic, semantic) do not hold up to the facts. But this is perhaps because linguistic behavior, assuming a handling of universals, is based on group scannings of recognition and neural mechanisms that are not organized in a way that corresponds to conscious impressions or to the conceptual analysis of language.[24] Head regretted not having the boldness to speak, regarding aphasia, of group x of functions, which would be subdivided into "groups a, b, c, d."[25] By trying to reproduce in automata the various group scannings of which Wiener speaks, for the recognition and handling of language, cybernetics would perhaps allow a mechanical analysis that would reveal itself to be deeper than any psychological one.

Without going as far as this, it would in any case be possible to produce mechanical animals that would closely mimic instinctive behaviour according to the three factors that the experiments of animal psychologists have highlighted: awareness threshold, action according to a thematic melody, and finally—which interests us in particular here—guidance by the perception of specific universals. It would certainly be difficult to imitate mechanically the conceptualized perception of human beings, and to produce a machine that, made to flee from snakes, would not flee from a slowworm, or that, made to hunt birds, would not hunt bats. But it would not be difficult to achieve the degree of perceptive finesse of most animal instincts, triggered or awakened by gnosis of an often very basic character, as shown by the effectiveness of bait and lures.

Nevertheless, cybernetics would be wrong to declare victory too early, and to believe that the problem of the perception of forms has been solved. The perception of an object as a "universal" presupposes in any case that hypothetical group scanning is capable of activating, at a certain moment, a mnemonic model. Cybernetics can only conceive of this model as materially existing, either in a certain number of determined cells, or, according to Karl S. Lashley's very vague conception,[26] throughout the cortex, where it spreads like a wave pattern. However, this activation, in turn, cannot be said to occur when the "mind" [*esprit*] recognizes the similarity of the mnemonic model and the current perception, once duly transformed, because nothing would be gained by reducing the problem in this way. The "mind" is here only a word or *deus ex machina*, which it is the very essence of cybernetics to reject. It will be necessary to admit a resonance similar to physical resonances, both to explain the formation of the schema—the universal—and to explain its reactivation. For example, for me to recognize triangularity in a particular triangle, a model must first have been formed from the previously perceived triangles, by eliminating the elements that do not reinforce each other: the values of the angles, the dimensions of the sides. Then, it is necessary that the new triangle reactivates the "triangularity" model.

Such a hypothesis is extremely unsatisfactory. Not to mention the difficulty of applying it to temporal universals (a dance or melody), or to things that cannot be reduced to a structural schema (for example, a "friendly attitude"), it is quite beside the psychological reality. When I recognize a plate, even when viewed from the side, the plate remains an elongated ellipse or trapezium, although I know it is round. According to the hypothesis, the final consciousness—after the intervention of the group scanning* mechanisms and the resonance with the cerebral model—should be that of a round plate. The immediate experience reveals a kind of combination of a very particular character where the sensory image is similarly transfigured by knowledge, but is not transformed, strictly speaking. It is a serious issue that cybernetics is thus led to disqualify the testimony of consciousness to such an extent. We can admit that the emotional atmosphere, or even the atmosphere of meaning, is not always essential in behavior. We can also admit that even the psychological interpretations of gnosis or agnosia are sometimes superficial. But in the perception of the plate, triangle, or hutch, it is difficult to reject an immediate intuition. We see the particular triangle, at the same time and indissolubly, as triangle and as particular. A resonance does not make the "three angles" appear, by blurring everything else. We see the plate as a circle seen obliquely, not as a circle or as a straightened ellipse. The study of what psychologists call the constants of perception reveals that the atmosphere of meaning, at least in this particular case, is not a vague epiphenomenon, but an effective component of perception, which obeys precise laws.

The least we can say is that it is really too early to declare, with F.S.C. Northrop, that the machines for perceiving universals have a "revolutionary significance for natural science, moral as well as natural philosophy, and for one's theory of the normative factor in law, politics, religion, and the social sciences," that they allow us to "overcome the dualism of mind and matter," of fact and value.[27] The Gestaltists had already been too eager to announce that they had resolved the same question.[28]

THE BALANCE OF INTERESTS AND COMPENSATORY ACTIONS

When a dog wants to grab a piece of meat, but is afraid of its master's stick, the stick becomes an obstacle that needs to be avoided. Or the animal can choose a less dangerous way to achieve its goal. An internal obstacle plays a similar role to that of an external obstacle. If the dog has a paralyzed or painful leg, it can keep moving on the other three, or press only lightly on the ground with the sensitive leg. Grey Walter's tortoise does not really take the obstacle into account; it does not really balance the obstacle against the "goal." Contact with the obstacle closes a circuit whose action momentarily replaces that of the phototropic circuit. To truly imitate the balance between two possible actions, not only feedback is needed, but a system of compound feedback. Dr. Ashby's homeostat is a device made up of four identical elements, each of which reacts to the other three.[29] One element has a movable galvanometer that controls the plunging of a metal wire into a conductive container with a gradient of potential. As each galvanometer receives the output current of all the others (not to mention, of course, its own output current), the balance of each depends on the balance of the whole. If the experimenter disrupts one of the elements by blocking the needle of its galvanometer, for example, the rest of the homeostat adapts to this new situation; it seeks and finds a way to reach the "prescribed" position of equilibrium. Furthermore, before reaching the windings of an element, the output currents pass through selectors mounted in "tiered" functions, which themselves represent feedback. These suddenly modify the main feedback when, due to a mechanical obstacle introduced by the experimenter, it would tend to take an extreme position instead of seeking the optimal equilibrium. The selector searches for secondary feedbacks that are suitable for the main feedback to accomplish its "mission," and for the device to search for and achieve the prescribed equilibrium again.

Ashby's device is capable of truly extraordinary feats. Suppose that the experimenter reverses the conductors of one of the feedbacks, so as to make it positive instead of negative, and makes it work like a reverse-acting ball regulator that would accelerate or stop the steam engine, instead of adjusting its speed. The selector then intervenes on its own to look for the feedback that will correct or switch off the reverse feedback and restore balance. If the experimenter connects the needles of two galvanometers with a rigid rod, the apparatus is also capable of restoring stable equilibrium, to the point that when the connecting rod is removed, the apparatus must again grope around to find the previous assembly [*montage*]. This is a little like a man who is healed but is disoriented because he had become used to his disease.

There are corresponding operations in the physiological order. If a dog's two left legs are connected with a rigid rod, it will be able to walk again, although with difficulty. If we surgically switch the internal and external straight muscles of the eyeball in a monkey, as in Marina's already old experiment, or if we switch the flexor and the extensor of the animal's arm, like Roger Sperry, the normal movements are restored—very quickly for the eye, more slowly for the arm—after a period of incoordination. Kurt Koffka made a big deal of Marina's experiment.[30] He saw in it proof of the existence of a regulating dynamic field, independent of the nervous "rails." Kurt Goldstein considers similar cases of compensation after an injury or deficiency to be evidence of his organicism [*Ganzheitlehre*], and of the independence of performance in relation to the activity of parts.[31] It must be admitted that Ashby's homeostat brings a truly sensational element to the debate.

His homeostat works by a set of conductors and not by a "free" dynamic field; and yet, he very precisely realizes the organic regulation that occurs after Marina's experiment. This regulation can now be interpreted as an automatic change of nervous feedback and not as the effect of a rebalancing of a cortical *Gestalt*. If the conflict between Koffka's interpretation and Ashby's does not appear to be very serious—after all, the actions and interactions of feedback are dynamic phenomena—it is above all against Goldstein's organicism and holism that the achievement of the homeostat seems to provide a decisive argument. The homeostat is obviously only a machine, and yet it acts as a whole that would be independent of its parts, of their damages or accidental constraints. We can easily imagine the philosophical developments of a [ganzheit-theorist or] organicist, a disciple of Goldstein, on the performances of Ashby's machine, if we tricked him into interpreting them while hiding that they are the actions of a machine. He would obviously see a *Tendanz zum geordneten Verhalten*, with *Auschaltung von Defekten*.[32] He would point out how "every change in one place implies a simultaneous change in another place," and how *die verschiedenenen Veranderungen sind ein Einheit*.[33] We certainly do not claim that Ashby's mechanistic theory is fundamentally true, but we recognize that cybernetics here scores a point against anti-mechanistic theories.

Ashby considers his device to be an analogue of the couple formed by the brain and the environment, the brain having to adapt to accidental and unpredictable variations in the environment.[34] But it can also be considered as capable of seeking a balance of interests. The automaton seems to calculate its behavior in the manner of a hedonist, or according to the formulas of marginalist theory. Undoubtedly, for the moment, homeostatic automata do not make real utilitarian calculations. Despite their groping around, their adaptation to circumstances is instantaneous, or at least "presentist"; we mean that they cannot plan the outcome of their possible action before they accomplish it. They only do physical experiments, not "mental" ones. But it is not excluded that it would be possible to construct some super-composed automata of which the homeostat would only represent one part, equivalent to the "speculating" cortex, the other part then using the results tested by the homeostat. The operations of seeking equilibrium and reconciling interests would first take place in miniature before being carried out on a large scale. Some already existing industrial processes can give us an idea of what this "mechanical mental experiment," if we can associate these words like this, would look like: Vauban locks, in Strasbourg, controlled by hydroelectric boxes where the pressures, on both sides of the valves, are reproduced on a metallic membrane; and the Network analyzer,* a reduced model of all the connections of an electrical distribution network, which makes it possible to study the various problems of stability posed by accidents or unexpected demands in the power supply.[35]

CONDITIONING AND LEARNING

Can an automaton be conditioned? Can it learn to take into account experience, so as to behave not only according to immediate results but, in the same way as Pavlov's dogs, according to the *previous* results obtained? Could it, for example, learn to react to sound, after having been set up [*monté*] to react to light, if sound were to replace light as a signal relative to the intended purpose? All the cyberneticians are confidently addressing this important issue. To fully understand the difficulty of the problem they want to solve, it is necessary to understand that learning does not consist in behaving, even with flexibility, according to a feedback system. Learning is essentially about setting up a new feedback system. What is learning to drive a car? It is essentially to set up cerebral feedback that is opticomotor (for handling the steering wheel), and audiomotor (for using the accelerator and gear stick). Novice drivers flood their engine or let it stall, because they want to use a more "natural" feedback, based on kinesthetic sensations, sensations that are insufficiently precise to inform the foot position accurately through a recurrent circuit. When the auditory information about the engine speed directly controls the position of the foot on the accelerator, the learning is normally completed. However, as we have seen, an automaton could be set up to drive a car. A radar, or photoelectric cells, could inform the steering effectors; an acoustic device could inform the acceleration effectors, or tiered effectors, for changes of speed. Moreover, as we have seen, like a good driver the automaton could switch from one automobile to another, whose steering is more or less augmented, or the accelerator more or less sensitive, since it would guide itself by the result obtained. But we must be careful of an ambivalence. This "transfer," from one car to another, would not be a real transfer, a learning experience. The automaton would not learn to drive: it would be set up for this performance, but it would not set itself up. It would not constitute the appropriate feedback for itself, like the human learner; it would not transpose the kinesthetic-motor feedback into a more effective audiomotor feedback. It is the modification or improvisation of an assembly [*montage*] through practice, and not the functioning of an assembly—even allowing for improvised effects—that constitutes learning.

It is true that Ashby's homeostat really seems to answer the problem. It modifies the assembly of its own feedback when it seeks its goal *despite* having a defective assembly to begin with. Ashby considers that his machine can explain why, in living species, feedback is set up correctly and not the wrong way: natural selection had to eliminate, for example, a species in which lowering the body temperature would trigger a further-cooling perspiration instead of a warming shiver. It can also explain how individual experience can correct feedback that would not be appropriate in a certain context, particularly by changing its sign (positive or negative). A child may be compelled to learn to look for a red blanket but to avoid a red ember. The details of this fine-tuning cannot depend on the action of natural selection, because different children, in different circumstances, can be adapted to different feedback arrangements. The homeostat can precisely correct feedback that is set up the wrong way.

But what Dr. Ashby forgets is that the uniselectors of his machine only choose between the assemblies inherent in the circuits. They do improvise a new behavior and equilibrium, but they do not improvise a new assembly. If the experimenter has reversed a feedback, "the uniselector then changes this feedback for another, by searching for one after another, among the randomly arranged feedback, until it finds one that fits. Once found, this feedback is retained."[36] We know that in conditioning experiences, on the contrary, anything can be used as a conditional stimulus. When the dog learns to salivate on the beat of a metronome or on the application of an electrical current to its paw, it is difficult to believe that a brain selector could have just chosen between "nervous feedbacks prone to chance." To this objection one could reply that in the cortex, there is a large enough number of neurons and conductors that the selection can always seem improvised. But then all the weight of the difficulty is transferred to the biological organization, which did not simply have to choose between ready-made assemblies.

Beyond the ordinary arsenal of cybernetics, other mechanical images of conditioning can be conceived, and have been attempted for a very long time. Let us admit—despite the increasingly pressing objections of most psychologists, which are based on experiments—that the neurology of conditioning can be schematized in the following way:

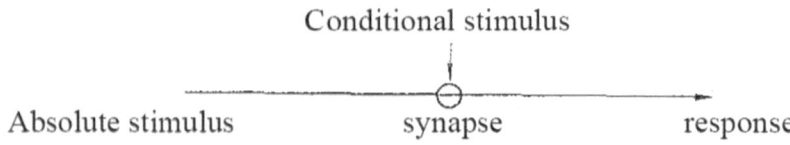

Figure 1.3.

Norbert Wiener believes that we can discern a mechanical model of learning.[37] When establishing conditional salivation, the important point is that the animal is hungry, a state that certainly has a physiological correspondence, for example a greater permeability of the synapses. Hunger is a kind of generalized internal state of alert, which changes physiological thresholds, according to a type of message or information whose social equivalent is provided by a bell ringer or alarm siren: "Warning to everyone the alarm concerns." If *any* stimulus occurs at the same time as the absolute stimulus, it can both take advantage of the greater permeability of the synapses and reinforce it by accentuating it, through a spawning effect [or in accordance with chronaxies]. This modification therefore amounts to a change in the assembly of the machine. We will not insist on this conception of Wiener's, which, moreover, only repeats very old hypotheses. Unfortunately for it, experimental psychology has shown that the true schema of the conditional reflex is very different. The two stimuli are not simultaneous, and above all the conditional stimulus does not provoke the same response as the absolute stimulus, but a cortical-based preparatory response (whereas the absolute reflex is a subcortical process). The conditional reflex looks like a voluntary action. B. F. Skinner has demonstrated that learning by trial and error and the action of what Edward Thorndike calls the "law of effect," that is, the stamping in* of the act that leads to satisfaction, would be exactly equivalent to Pavlov's conditioning. In "Skinner's box," the rat quickly learns to approach the food when it hears the sound of the meatball that the experimenter drops on the tin plate. The sequence is exactly the same in both cases:

Pavlov: metronome—salivate—eat

Skinner: sound of the plate approach—eat

The only difference is that "approaching" is more directly instrumental than "salivating." If the rat accidentally presses the bar that drops the food pellet, the sequence becomes Bar—Press—Sound—Approach—Eat.

It is clear that a simple self-regulating automaton only appears to imitate the "law of effect." The thermostat seems to be looking for a given temperature. Once this temperature is reached, it seems "satisfied," since it maintains it. A D.C.A. cannon, with radar, also looks for its objective until it has obtained the "satisfactory" effect. But for one to be able to talk about learning or conditioning, this "satisfaction" would have to concern the means and not just the goal. The living being knows how to find the most expedient means. "Pressing the bar," for the rat, is an act "valued" through success.

A secondary automation must then necessarily be superimposed on the primary automation so that the machine imitates the living being. This secondary automation would itself work on information, but on statistical information about successful moves; in short, it would be able to induce. However, as we have seen, induction machines, without being inconceivable, are very difficult to realize. Not to say that it is extremely doubtful that this would be authentic induction and therefore true learning.

The problem might be a little less difficult if it were not a question of "satisfaction" and induction, but of a simple application of what Thorndike calls the "law of exercise" (the influence of repetition, "recency," and "frequency"), a law which is in fact far from being unanimously accepted today. Wiener again, probably not satisfied with the aforementioned hypothesis, and relying on recent research from the Eindhoven laboratories of the Philips Company,[38] gives a general idea of what automatic learning based on this law could be. If I call a colleague or a close friend twenty times a day with an automatic telephone, the procedure for the call is just as complicated as that for calling an occasional correspondent. A living being in a similar situation would soon find faster alternatives for the often-repeated action. It would be necessary then that the apparatus of the automatic telephone records the relative frequency of the numbers called, and modifies its own assembly, through second-degree feedback, for the "most frequent calls." We appreciate the extreme difficulty of mechanically making a device that modifies itself in the appropriate sense, other than by simple mechanical wear. Simple wear can sometimes be equivalent to improvement, but the running-in of an engine, the softening of new shoes, or the increase in capacity of a battery after use, cannot be considered to be authentic cases of learning.

However, cybernetics does not give up on tackling the problem even in cases where induction is necessary. Take again the D.C.A. cannon with automatic aim. To avoid projectiles, the targeted aviator will modify his line of flight, but he cannot modify it *ad libitum*: he is dependent on the centrifugal force and the characteristics of the aircraft, which prevent him from making very sudden turns. A human aimer would get used to these evasive moves and would aim their shots accordingly. However, a device can also do this by feedback of the second degree, superimposed on the automatic functioning of the first degree, and functioning with statistical information of the most common evasive maneuvers of the aircraft.[39]

[Another conceivable method of approach for the problem of learning has been attempted. It starts from the fact that a negative feedback can be more or less efficient, that is, it can stabilize its oscillations more or less quickly. Certain systems can only be stabilized following their task by at least two feedbacks, not only one. An auxiliary feedback can test the characteristics that it would be necessary for the principal feedback to have in order to render the system stable in a given circumstance. This is equivalent to a sort of learning.* One can take for example the way we behave when we walk or drive on an icy road.[40] All our ways of walking or driving depend on our knowledge of the degree of adherence of the road, or in other words, of some "oscillatory characteristics" of the pedestrian-road system or car-road system. If we drive with the authority of our usual manner, we risk going into the ditch. So while driving we make a series of small steering movements, insufficient to make the vehicle skid dangerously, but sufficient for us to be continually informed of how the vehicle is responding. We then adapt the way we drive in response. A "feedback informing a feedback" can be added to an automaton. It is, however, doubtful that it is a matter of true learning here. It is only a matter of an automatic choice between two preexisting assemblies.

We have thought of still other methods, in particular those that equip Grey Walter's tortoise with the capacity to learn. Namely, the learning represented by the well-known process: "The burnt hand is afraid of the fire." We can put a fuse on the approach circuit of the tortoise. If it makes contact with a burning object, the fuse melts and breaks the circuit of the approach mechanism, and another, backup mechanism comes into action. If the tortoise crashes into the burning object too often and too violently, the effects of the shocks can be accumulated, for example on an irreversible spring mechanism, which eventually triggers a device for bypassing the obstacles without touching them. However, it is apparent that in these methods, the learning* is "cheating." They only work once and are anticipated by the constructor. They are "cheating" like the pseudo-intelligent acts of a trained animal.[41]

Shannon, among many others, is currently trying to solve the problem of the automatic telephone's learning to make more frequent calls. To begin, he has very recently focused on an "automatic mouse" that "learns" a maze. It first wanders the maze, groping around, and only arrives at the end after a lot of trial and error. But since the device is endowed with a memory, if we then put the mouse back at the starting point, it arrives at the end without error. Before Shannon, Albert Ducrocq had made a similar device. The problem of automatic learning is thus apparently solved. But this is only apparent. The mouse is guided by a control mechanism, with which a preestablished program conducts a systematic exploration of the maze, which is divided into squares. An automatic memory records the success in each of the squares, and the order of these squares. This memory can then serve as a program for the following performance of the device, which guides the magnetic mouse without error. It is obviously not a question of learning, but of mechanical memory, the possibility of which is not denied by anyone. Doubtless we will soon achieve other impressive imitations of learning, as we have achieved impressive imitations of perception. But imitation alone is not interesting. What would count would be an authentic "model" of the phenomenon.

The theory of learning is, along with the theory of universals,* the weakest part of cybernetics. Experimental psychology has demonstrated that learning clearly involves something other than the implementation of functioning, or even the assembly of new nervous circuits, or the reorganization of the perceptual field according to the laws of common (that is, macroscopic) dynamics. Animal learning, as has been demonstrated, notably by E. S. Russell,[42] always takes place on the ground of an instinctive *Umwelt* and is part of an oriented and "conative" activity, which aims at the satisfaction of a need. The animal puts the accent of "value" or "importance" on some detail that it first ignored. "Success, however achieved, directs the animal's attention to the significant features of the perceptual field, those which are significant as means to the end pursued. "[T]he perceptual field is organized with respect to these significant features, and when the same situation is presented again the field retains its special organization—the animal sees it again in this special way. It is therefore able to repeat the solution, for it now perceives the situation in the light of its first success."[43] In short, cybernetics misses the point of learning as it does of the universal, because in both cases it is impossible to imitate meaning mechanically.

On the other hand, finally, the interpretations of learning proposed by Wiener would strictly speaking only take into account reflexes learned "in response," and not "operant" reflexes, to use Skinner's terminology. With the automaton, the response that is learned would always be consistent with the response that is not learned. But the animal often changes its method in the process of learning; it attempts some different operations when the first have failed. For example, the cat pulls instead of pushing; the monkey that can no longer find the stick tries to break the bars of the cage or begs the caretaker. From this point of view, the schemas of Grey Walter would appear to have the advantage over Wiener's. This is because the automata, once perfected with some filing mechanisms, moves from one process to another, by putting a process that has failed out of operation. But this advantage is only apparent. In an electrical installation, we can easily predict the automatic starting of an auxiliary motor in the case of the primary motor breaking down. No one, however, would be tempted to speak of the learning of an electric power station. In the Nividic lighthouse, which does not have a caretaker, a siren automatically replaces the light in the case of fog, and a flare gun automatically replaces the siren if it stops sounding.[44] But no one speaks of learning in this case, any more than they would for the automatic telephone message "Please refer to the new telephone directory." Truth be told, the cyberneticians could respond that an organism also possesses a limited repertoire of possible operations: the cat can only chew, scratch, push, pull, or meow. Indeed, animal learning usually consists of calling successively on virtual activities that a state of alert sets in motion. But for man, at least, the new operations possible are indefinite in number. Let us consider, for example, the variety of means a person who is cold can use to warm themselves: go for a run, put on some warmer clothes, go to bed, buy an electric radiator, do Swedish gymnastics, prepare some hot tea, drink alcohol, go to the cinema, take the subway, eat more food, go to the South, use the Coué method (auto-suggestion), and so forth.]

Chapter 2

Framing Activities and Framed Mechanisms

[In the last chapter, we saw that all the internal difficulties of cybernetics stem from the same error of principle and the biased postulate that information machines are the complete equivalent of living and conscious nervous systems. This mechanistic postulate drives all the failures of cybernetics: the failure to understand the origin of information, and the implicit acceptance of a true perpetual motion of the third kind; the failure to understand meaning; and the failure to understand the perception of universals,* or learning.* We must therefore move on to a deeper and more general criticism, or more precisely, a positive reinterpretation of cybernetics free of its mechanistic assumptions.] We must give up the claim to replace the conscious nervous system with machines and instead consider the machines as subordinate to the living nervous system and framed by them. First and foremost, we must acknowledge that the assembly (in the active sense of the word) of any mechanism is quite different to the assembly (in the passive sense) of the fully constituted and functioning mechanism. Active assembly is the work of consciousness, which creates connections according to a meaning. Passive assembly is the set of connections once they are restored, and automatic assembly can replace connections improvised by consciousness.

What is the common element of all automatic assemblies? What does the constructor spontaneously look for when he wants to mechanically imitate conscious behavior? The answer lies in this word: connection [*liaison*]. The manufacturer is always asking himself, "What kind of connection should be made to achieve the desired effect?" For the calculating machine, this is obvious. The construction problem, for the radio-handyman who wants to make a "Simon," as well as for the professionals who want to build a new Mark, it is a problem of connection. The manufacturer obtains hundreds of electronic tubes and hundreds of meters of wire, and starts welding according to the assembly diagram. Kalin and Burckhardt's machine existed in principle from the day Shannon realized that logical algebra was translatable into electrical circuits. For self-regulating automata, this is no less obvious. It should only be remembered that there may be connections of a different kind than sliding linkages [*liaisons par glissières*] or electrical wires. An electrical or magnetic field, or an ordinary dynamic constraint, can be used as well as a kinematic connection. The functioning of the regulator—that of Watt, for example—depends on the establishment of a connection that is on the one hand kinematic (the regulator deforms and acts mechanically on the intake), and on the other hand dynamic (the deformation is carried out by centrifugal force).

But before the assembly was carried out, before the purchase of wires and welding equipment, the connections had to exist already in the constructor's consciousness. A machine with imperfect regulation must be guided by a supervisor, whose consciousness "closes," in some manner, the circuit of a semi-mechanical, semi-psychic feedback. The machine yet to be built exists as a semi-differentiated schema, the parts of which are linked by the general meaning of the effect to be obtained. Before this schema, an even more thematic premonition existed in inventive thinking. From the prescient consciousness to the schematizing and surveying consciousness, then to the supervising consciousness that completes the machine, we finally arrive, with full automation, at the total replacement of consciousness by a set of "substituted connections."[1] These reproduce, by pushing or pulling, by conducting from one person to another, the primitive connections inherent in the "absolute survey" that characterizes consciousness. Consciousness, or meaning, is not an ineffective atmosphere; it is—if we can oppose metaphor to metaphor—a "creative nebula," of which the automaton is only the residue. The feeling of "and" and "or" is transformed, for example, into a pattern of "two paths placed end to end" or "two equivalent paths." This schema is in turn transformed into real electrical circuits. The feeling: "The machine is going too fast, the intake should be reduced," changes into "The speed should control the intake," then into "Automatic control schema," then into "Actual control by negative feedback."

The metaphor of "nebulous consciousness" is still only a metaphor. But cybernetics itself suggests a more precise definition. Consciousness without auxiliary machines is a kind of putting-in-circuit of the elusive x-center of individuality with the world of meanings and values. Consciousness implies a kind of axiological feedback, which is unrepresentable in its totality in the spatiotemporal world, through which trans-spatial meanings and values inform, by recurrence, the spatial part of the nervous circuits. Anticipating consciousness is in contact with an invisible and unobservable, though not supernatural, meaning or ideal that controls its processes of realization. The "ideal," which can be materialized in the automatic machine, is obviously second to the nonmaterialized ideal of the conscious being. Similarly, "substituted connections" are second in relation to the connections improvised by consciousness, or the actual assembly in relation to the designed assembly. It's because I want to get warmer that I turn up the thermostat. Machine automation is a kind of spatial projection of something that is hyper-spatial.

[This projection is always imperfect and incomplete. In many cases, it is even totally impossible. When my hand, monitored by my visual field, goes to take a full glass on the table, the ideal meaning of "To take the glass without spilling it" is almost perfectly replaceable by an "ideal" for an automaton, controlled by photoelectric cells, and operating by means of stabilizing mechanisms. The "surveillance" by the visual field is not really replaceable, but its effects are imitable. But when the hand is that of an artist seeking a harmonious line, the invisible ideal, which nevertheless compels him to erase imperfect lines through a kind of negative feedback, cannot be replaced by any mechanical substitute. We can conceive of machines for making pointillist or cubist paintings starting from an automatic "perception"—and some artists' brains seem to have worked as such a machine would work—but art would not be well served by it. How can we mechanically represent the regulation exercised on behavior by a demanding "superego," and above all by a purely spiritual ideal? Even in the case of the hand searching for the glass, there is much more in consciousness than in substituted automation.]The visual field, with its multiple details, supervises and surveys itself, in a unity that does not imply the existence of an external surveillance point. From such a field, by definition, all kinds of connections can be improvised and realized, because they are already virtually present in its *unitas multiplex*.[2] Moreover, the passage from consciousness to the machine is always possible. There is no *more* in the machine, but *less*. The machine is an extract. There is less in the automatic connection, or in the materialized ideal, than in the consciousness of the person who provided the connection in a flexible and improvised way, according to an invisible ideal. The hand and the eye, the nervous conductors that guide them, and the observable parts of the cortex that control their operation by means of nervous feedback, are themselves already auxiliary machines for the direct dynamic action of meaningful consciousness. The assembly of the mechanical automation is preceded by cerebral assemblies, themselves controlled by the binding action of the intention aiming at a hyper-spatial meaning, and informing the perceptual field or the behavioral field according to this meaning. If I am in a hurry, an essential value being at stake, first my psyche, then my nervous system are set up according to the theme "Maximum speed according to the circumstances." Eventually, an auxiliary machine, such as a bicycle or automobile, is put into operation to achieve this maximum speed. Consciousness envelops the brain assemblies as well as the auxiliary mechanical assemblies that complement or replace the brain assemblies, ready to correct both. Between the feedback of conscious connections and mechanical feedback, the brain is the domain of mixed feedback, the place where active assemblies are transformed into passive ones. It is a flexible and unfinished machine, easily and temporarily closed according to changing intentions.

In relation to a machine whose automation is imperfect, the user's or worker's awareness plays the role of a pure auxiliary, framed by the ensemble of the mechanism, and with improvised connections making up for what the assembly of the machine leaves poorly connected. At this level, the human being "supplements without loss whatever human faculties the machine lacks."[3] For example, if a red light appears on a car's dashboard, indicating "Water too hot," I have to stop and remove the radiator cover that I had forgotten. An improvement would be the automatic adjustment of the radiator plug according to the temperature, eliminating the need for my conscious concern. We therefore understand the conviction of the cyberneticians. The improved automation perfectly replaces the conscious connections that remained inserted in the cycle of poorly adjusted mechanical connections. So, they conclude, automation represents the entirety of conscious connections. But we also see their mistake. The "inserted" consciousness, far from representing consciousness in general, represents only an incidental and accidental role. The essential thing is the framing, enveloping consciousness that had to invent and combine all the mechanisms of the car as a means of transport. The isolated zones where consciousness needs to intervene, which remain in cases of imperfect mechanical operation, are destined to be eliminated. But this does not mean that the framing zones can also be eliminated. A well-automated device can be operated without conscious concern, but it must be built with even more conscious care. Just because lakes are a geographically transient phenomenon, this does not mean that the oceans are destined to disappear in the same way.

THE "SET"

We will now examine the relations of the framing activity and the framed mechanisms more closely. Suppose that

> I {seek to achieve}[4] an organic state of comfort.

Outside the brackets is the framing part: the active individuality on the one hand, its ideal on the other. Within the brackets is the framed part, which can be more or less mechanized. The organic state of comfort is not a pure ideal in the sense that it does not have to be the object of a pure invention. I have already experienced it, so it is "mnemified," and it also depends on the current state of my organism. But, whether pure or impure, it nevertheless has the essential character of the authentic ideal, of not being physically inscribable in a device. Now let's insert a device.

Sitting in my office in the winter, I find the temperature a little low. {I go to my apartment's heating boiler, and I raise the graded scale set on the thermostat, from 50 to 60 degrees. The thermostat works, the temperature rises} and I feel more comfortable. Another example. I'm going to the place of a friend who lives on the fifth floor. When I get to his house, I set myself {to climb the stairs} to reach the required floor. If there is an elevator, I {press the "Fifth Floor" button, wait for the elevator to execute the "order"}, and enter my friend's home. An aviator, if he feels diverted from the initial course, or if the aircraft "rolls," {consults the artificial horizon or course indicator, then acts on the controls} and brings his aircraft back on course. If the aircraft is equipped with an automatic piloting system, the pilot does not even have to feel the deviation: the gyroscopes, with vertical and horizontal axes, control the maneuvers necessary to correct the deviations themselves, by means of pneumatic servomotors. The aviator only intervenes if the automata do not work, just as I take the stairs if the elevator breaks down, or I exercise to warm myself up if the heater doesn't work (see table 2.1.]).

We could indefinitely multiply the examples, drawing them from either an effort of memory using auxiliary mnemonic means, or from an effort of imitation, economy, expression, and so on.

We see that in phase 2, we can have either the activation of a fully assembled, effective machine or the assembly of what psychologists call a "set."* It has become established practice to use the English word "set" to refer to any psycho-physiological set-up, any neuromuscular adjustment that favors an action, either by preparing it (for example, "taking the physical and mental attitude for starting a race") or by favoring it through control while the operation is in progress (for example, "sacrificing everything for maximum speed during the race"). The set acts as a selective factor, facilitating some responses and inhibiting others. The set is the putting into relation of the neuro-psychological apparatus and a meaning or value, beyond raw sensation. Without the set, the sensation would remain in the state of ineffective impression and would even cease to be a conscious sensation. While driving a car, I cannot, without danger, be in the state of an impressionist painter. I watch out for vehicles coming out of the side streets, and any shadows that appear, given value in advance by the psychological assembly: "Beware of vehicles crossing my path" very quickly triggers the appropriate movements. Similarly, "Drive on the right" is valued in relation to "Drive on the left." What is performed by the material assembly of a device is always a valuation of this kind. A thermometer in a room, which does not control any temperature setting, is therefore the equivalent of the "impressionist" state. If it is inserted into a thermostat to regulate a heater, the indicated temperature is efficiently given value: it becomes the optimum temperature. Whereas the assembly of a machine is only given value by human intention, this intention, commanding the corresponding psychological set, must obviously be in direct relation with a meaning or value. The cerebral cortex, the place of the primary manifestations of the psychological sets, is related to two worlds: the world of meanings and values, the apperception of which is a constituent part of the set, and the world of our space, where it achieves the effect according to the set.

In most of our actions, we have both an "aim-assembly" (our psycho-physiological apparatus is set up according to the intended aim) and a "situation assembly" (it is set up according to the current situation). For example, if there is no elevator available, and the stairs are long and steep, I take a deep breath, and I adopt a rhythm that would not be that of a

Table 2.1.

1	2	3	4
I need to feel warm	I turn up the thermostat to 60°	The temperature of the heater rises	I feel warm

Or:

1	2	3	4
I Need to feel warm	I start exercising	the increased bloodflow warms up my skin	I feel warm

All human actions, without exception, can be represented in an analogous way. Here are some examples:

1	2	3	4
I ask which of two objects is heavier	I place them on a scale	the scale works	I find out which object is heavier

Or:

1	2	3	4
....	I just hand weigh each using the psychological comparison effect produced by weighing the first	the psycho-physiological comparison effect occurs	....

1	2	3	4
I'm looking for the product of two numbers	I use an algebraic rule	The rule gives a number	I know what the product is

1	2	3	4
I'm looking for a pattern for a decoration	I rotate a kaleidoscope	It forms a new pattern	I am inspired to draw the pattern

"hundred meters sprint." If I weigh two objects to compare their weight, it is the psycho-physiological set, maintained from the first to the second object, rather than the memory of the first, that allows me to make the comparison.[5] If I do exercises to warm up, the set is very different to what it would be if I aimed for economic productivity in my movements.

The ready-made assembly of a device, or more precisely, the last assembly, operated by the control button or lever, is the equivalent of the set. In the case of the set, organic machines are used instead of external machines. But always, even for the simplest living beings, those least different from the macromolecule, there is an intervention of machines in the performance of the intended act. In a car, if I want to go faster {I hit the gas, the engine accelerates}, my desire for speed is satisfied. In a running race, if the runner feels threatened by a rival near the finish line, his willingness to win, or his fear of being beaten, triggers an emotion—that is, physiologically, a hypersecretion of adrenaline—which accelerates muscular exertion. The will to win acts on the central nervous system, which acts on the sympathetic adrenal system, which in turn, by accumulating adrenaline in the blood, overcomes the effects of muscle fatigue. Emotion can be considered as a very general set, facilitating the passage from the will (or the spiritual apprehension of a value), to the more particular psycho-physiological set, which, like an operation on an auxiliary machine, allows the attainment of the goal.

The psycho-physiological set is of course already impossible to reduce to a purely mechanical assembly, despite the frequent possibility of a substitution. The motorist who sets himself up for the set "Maximum speed" makes the most diverse maneuvers according to the slopes, turns, and obstacles. It is difficult to conceive—although technological progress can achieve it asymptotically—an automation for the driving of a vehicle such that the driver only has to press a command button: "Maximum speed according to the circumstances," so that the machine obtains it mechanically and that between the pure will of the driver and its realization, there would be no psycho-physiological intermediary. But if the set is psychological relative to the subordinate mechanism, it is already semi-mechanical relative to will or pure intention. It works thematically but in a semi-blind way. The proof of this lies is the absurd perseverances it often causes. A psychological "task," once set up, continues to act, even when it no longer corresponds to an intention, like a machine that we have forgotten to turn off after we have finished using it. In most cases, instead of considering the set on the one hand, and the machine on the other, as interchangeable, it is better to consider them as framed within each other. This will give us the following:

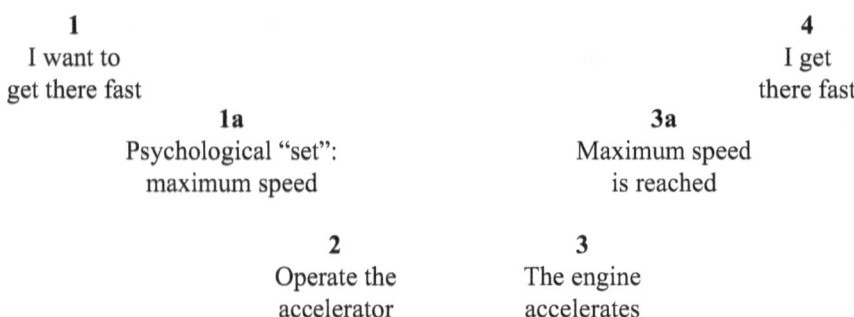

Figure 2.1.

In the case of organic development, we do not have the right to project ourselves empathetically in place of the organic x, or to say that the final state was the goal that this x aimed for. But the general schema is nonetheless exactly similar to that of conscious psychological action, in that there are also mechanisms, which are generally chemical, interposed between the beginning and the end of development. Some region, itself triggered, releases a hormone or an organizing substance, which produces effects on some other region—effects that seem to respond to an ideal that is aimed at. In the metamorphosis of amphibians, or in the puberty of mammals, thyroxine or sex hormones act on development, just as adrenaline acts on behavior in the case of emotion. The schema:

1	2	3	4
The organic x tends towards the adult state	Release of thyroxine	Action on sketched out organs	Adult state

Figure 2.2.

This schema is obviously metaphorical in that it speaks of a "tendency" or "willingness" at stage 1, but it is consistent with the facts in that it frames stages 2 and 3 by stages 1 and 4. For it is a fact that the "meaningful," if not intended, transformation of the young adult has the chemical triggers as "means." The proof lies in the fact that neighboring organizations often use very different means to achieve the same goal.

There may be actions that could be called direct, that is, without means. I can judge or reason intuitively, without using machines or slide rules, or even, perhaps, without using a psycho-physiological set. In fact, it is very difficult to confirm that a psycho-physiological set is not interposed. So-called "mental" calculation most probably uses some brain circuits as instruments, otherwise we would not understand why it is tiring. The practically universal case is that not only one, but several mechanical or psycho-physiological functions are nested in each other, as in a Chinese box.

SCIENCE AND THE FRAME

Science is much more at ease in explaining the mechanical or semi-mechanical median region of actions than in understanding their origin and end. Doctrines such as existentialism; psychological conceptions that address the "I," such as those of William Stern or Mary W. Calkins; and biological or psychological vitalist theories, are hardly welcomed. The same is true for Platonic or Husserlian philosophies of essence, for axiology, or for the theory of organic types. In short, science neglects everything that is "framing," whether it is the "agent" or the "ideal." When there are several nested "means," for example when the operation of an auxiliary device is itself nested in the half-operating action of the psychological set, this "law of negligence" is even more striking. The more "central" the means, the clearer and more scientifically accessible they are. There is nothing clearer than the operation of a thermostat, an elevator, a Directional Gyro Sperry, or a slide rule. The operation of a physiological set is already more difficult to understand. Even more difficult is the action of a psychological "task." The psychologists at the Würzburg school, like the psychoanalysts, have been accused of pseudoscience. Finally, the action or nature of the will, of the "I" or of the organic x, the intentional relationship of the "I" to an ideal, is quite mysterious. So we understand that science has always tried to interpret the framing elements, 1 and 4, starting from the framed elements, 2 and 3; the encompassing action starting from the auxiliary functioning encompassed. Cybernetics is only the latest, most precise, and most interesting attempt to reduce the whole "framing" to auxiliary functioning.

And yet, only instinctive or voluntary motivations give meaning and possibility to auxiliary assemblies. Isolated assemblies make no sense and are literally nothing. The automata of cybernetics, having no motivation, are not beings. Grey Walter's automata are pure passages from stimulus to response, while real beings go toward their goal by way of means and obstacles. Automata actually operate according to a framing will: that of their constructor. They are auxiliary machines for the constructor like all man-made machines, and it is absurd to use the auxiliary as a model to understand the autonomous being.

The mechanical causal sequence from 2 to 3, and the causal and thematic sequence of the psychological set (which is at the same time a "push *a tergo*" and an "orientation toward an optimum") from 1a to 3a, imply the finalist action 1–4. The mechanical sequence is without consciousness, the action of the psychological set is subconscious, but the whole is set up by intentional consciousness.

[In the experiments with Dr. Ashby's homeostat, the set is represented by the various adjustments of the assemblies. Dr. Ashby makes a mistake of vocabulary or designation on this subject that appears unimportant but which is of great consequence for philosophical interpretation. He considers the interventions of the experimenter as representing the "modifications of the environment," and the rest of the apparatus as representing the brain that has to adapt to these modifications.[6] This is a serious confusion. The environment and its changes often impose a "task" on us (for example, the lowering of the temperature imposes on us the task of raising the thermostat's thermometer). But it is the living being himself who must "rise to the task" in order to respond to the change. The brain is therefore represented by the entire homeostat, including various assemblies, and not only by the part of the apparatus that responds to the constraints that the experimenter imposes on the other part. Consciousness is represented by the intervening experimenter, which means that it is not explained at all by the apparatus. It is precisely the action of *taking on* the task imposed by the environment. Taking on a task, through a set, is not the same as being subjected to it.]

THE AXIOLOGICAL FRAME

[It is remarkable, however, that the most positivist value theorists—such as Stephen G. Pepper,[7] who claims to be a follower of Ralph Barton Perry and Edward C. Tolman—are led to schemas very similar to those we have proposed. In order to describe a real activity—for example that of the aviator who, after falling into the sea, is hungry in his dinghy, and sets out to catch fish—Pepper insists on the need to clearly distinguish the "governing propensity"* (here hunger that seeks to achieve a state of appeasement, a quiescence pattern*) and subordinate acts (here fishing) tending toward a defined goal-object (the fish as food). This gives the schema:[8]

Governing propensity		Subordinate acts	Goal	
Tendency	Anticipating set		Goal-object	State of appeasement

Figure 2.3.

The operation of the hook or net can itself be regulated in "subordinate acts." However, according to Pepper, the "governing propensity" is clearly the primary dynamic factor in the whole structure of action. The subordinate acts (and, we will add, the mechanical operations used) are subordinate to the "anticipatory set," which itself depends on the drive,* just as the goal-object* is subordinate to the quiescence pattern, and the value of the fish, as an object, is subordinate to the value of "appeasing hunger."

A machine can imitate "subordinate acts"; it can even contain, as we have seen, the approximate equivalent of the anticipatory set, and reach the goal-object by itself. But it is inconceivable that a machine encompasses the governing propensity and the quiescence pattern and that it is an autonomous source of value or of value awareness. And yet it is quite clear that value and awareness come first here in relation to framed operations.] Let us try to adopt the hypothesis of cybernetics and consider man acting as a pure automatic machine. What do we gain from this hypothesis? We simply push back the problem by making it more acute. If the man who, desiring to be warmer, raises the thermometer of the boiler, is still himself like a mechanical device; if the desire "to be warm" is materially inscribed in his brain, as the temperature he desires is materially inscribed in the device, and if this desire is nothing other than this material inscription—then by whom is it written? By which super-engineer or super-user? If the engineer is the same, in principle, as his automata, then who is the engineer who made the engineer? We do not claim that there is not, in the brain, at the moment when it performs a defined task, a material arrangement regulating the action of its nervous feedback. But the active establishment of the set is necessarily a characteristic of the real being of which the visible brain is the appearance or the organ. If the action itself, 1–4, is in reality only a mechanical operation, it will be necessary to have recourse to a God, or rather to a Demiurge, childishly imagined as Engineer and Manufacturer. If on the contrary it is accepted, in accordance with the facts, that each of the nesting systems differs qualitatively from the nested systems, that the psychological set is different from the mechanical assembly, and that the conscious intention in turn differs from the psychological set, then we will have a means of glimpsing, through a kind of qualitative extrapolation, the nature of what frames the activity of man and of all living beings.

In any case, there must be a frame, 0–5, which may itself be complex, for the whole system of human or organic activity. The animal or man is not an absolute beginning. Tendencies or wills, oriented by valences or values, which contain subordinate functions, are themselves enveloped contents. Living beings are not the fabrications of a Demiurge, or a Great Engineer, who would endow them with a certain number of sensitivities or "variable needs," in the same way that the engineer arbitrarily gives to his automaton "sensitivity" to light or to the charge level of its battery. But in a more subtle sense living beings are "creatures," "axiological automata," created and "framed" by a Transcendent who fixes their nature and the values to which they are sensitive. These values or valences control their actions by a kind of axiological feedback similar, but not reducible, to the mechanical feedback of automata. Something in animal instincts is imitable by machines. It is possible to make artificial animals that flee from the sight of a cane, like a dog; that head for a heat source, like a bug; that swim against the current, like a migratory fish; that walk in a line, like processionary caterpillars. But obviously, something in divine art is more subtle than the art of the engineer. The valences that drive instincts, and especially the values that drive human behavior, are not reducible to physiological, or even psychological, mechanical assemblies, which are there to process them. When a man sets himself an elevated goal, an ideal difficult to achieve in art, morality, or technics, his actions are also controlled at every moment according to the results already achieved, by comparison with the ideal sought. Basically, the situation is the same as when Elmer and Elsie, Grey Walter's automatic tortoises, move toward moderate light. But it would be too paradoxical to reduce axiological feedback to mechanical feedback. Living beings are at the same time in physical space and in an axiological space. Psycho-physiological assemblies (and already brains, as observable apparatuses) are only projections on geometric space of a reality that overflows them. Visible feedback is only a "degenerate"—in the sense that physicists use this word—state of axiological feedback. As is already the case with an animal disobeying its instincts, a man who disobeys the ideal that usually keeps him in check is worried, troubled, and dissatisfied. This dissatisfaction can hardly be considered as the epiphenomenon of the oscillatory movements that bring the behavioral effectors back to normal equilibrium. The discomfort one feels when leaving one's axiological path can hardly be equated to the pounding felt under the soles of the feet when one has left one's geodesic space-time and feels "lead-footed." The strength of an ideal is not reducible to force in the physical sense. It is even the exact opposite: force in the physical sense is only a statistical appearance of elementary axiological forces.[9] As often happens in the psychological order, it is the higher-order action that must give the model of explanation for the elementary action, and not the other way around. An artist who lacks the taste to judge himself and his own productions could not progress toward his aesthetic ideal. Taste, or value judgment in general, has exactly the same function in higher acts as sensory information in elementary acts. An "ethically colorblind" person, a man suffering from axiological blindness, doesn't know how to regulate his actions, just as a blind person or a tabetic doesn't know how to regulate his movements.

It is in another, trans-spatial world that man, and already animals, seeks regulatory information. It is impossible to conceive of a machine capable of regulating and informing itself by value judgment. We can metaphorically, in a mind game, imagine all the efforts of humanity in all areas, throughout its history, as a vast set of feedback actions, in which invisible (and indeed poorly defined and changing) ideals control the already achieved results at every moment. This metaphor is not without value, but it is obviously only a metaphor. "Regulation by value" is something more than a mechanical regulation. People who fear the mechanization of humanity through technology seem to believe that, for example, automobiles, through the power of improvement, will first have automatic steering, then will be able to follow a road on their own according to a program, allowing the owner to stay at home while their car travels. They will then be able to choose their own route, according to the roads indicated as scenic by a guide; then they will be able to explore the roads and determine which are scenic by themselves. These fears are entirely childish, precisely because, if axiological feedback is similar to mechanical feedback, it is also essentially different, and above all it envelops the latter. Something transcendent, in man and beyond man, will always frame his industrial machines or his physiological machines.

Chapter 3

The Space of Behavior and Axiological "Space"

The most shocking thing about nonmechanical postulates is the invocation of a "trans-spatial." Because of an ingrained prejudice, "trans-spatial" seems to be equivalent to "supernatural." This prejudice is all the more unjustifiable because in physics itself, theorists increasingly use "configuration spaces," which are beyond ordinary space. It is quite possible that some parts of the trans-spatial world may blend seamlessly into the trans-natural. But, as far as the area we are interested in is concerned, the trans-spatial must be considered as a perfectly positive hypothesis.

Since the question is of utmost importance, we must insist on it and try to clarify it with schemas. According to orthodox cyberneticians, as well as gestaltists, we must be able to depict any feedback and any circular process, or dynamic regulation of a gestalt, or homeostatic regulation—we have noted the close kinship of all these concepts—without leaving the space-time of physicists, or at least while remaining in a "behavioral field" (Koffka) or a life space* (Lewin). While having certain particular characteristics contrasting it with "geographical space" and the physicists' space-time, the behavioral field or life space has in common with the latter at least the fact that it does not call upon anything transcendent, and that time only appears there as a simple dimension, the instantaneous slices of life space being always dynamically sufficient to themselves. The field of behavior, "life space" in this conception, has immediate dynamic properties. "History" only intervenes mediately, through its current dynamic effects. The same applies to intentions bearing on the future. Animal instincts, like human ideals, are neither innate nor transcendent. Apart from external factors and their disruptive effects, continuously occurring changes in the field of behavior are due exclusively to the instantaneous dynamic interactions of its parts. When the organization of these parts presents disequilibria, the field is unstable, and tensions arise. The tensions, in turn, provide potential energy for the work of readjustment. In short, the regulations, whether by feedback or by Gestalt and self-distribution, are not essentially different in the field of behavior or in the space of physical phenomena. In both cases, actions occur according to forces *present* in the field. If A, in the behavioral field, represents "me," or more precisely my hand, and B is a distant object with a "calling character," the distance of B from A creates a tension. This tension decreases if A and B come closer and increases in the opposite case. A and B will therefore tend in one way or another to come closer, according to the principle of least action. Of course, the way in which this decrease in tension takes place is complicated due to the fact that it is a matter of the space of behavior and not geographical space. No attraction occurs between A and B as physical objects. No object attracts my hand directly (a strong magnet only attracts it if it is holding an iron bar). The attraction is between A and B as images or psychological realities.

But this complication is no different from that of industrial appliances that have a guide mechanism, on which the general operation of the system is "dependent." Psychic attraction or the "calling character," according to the hypothesis, produces effects directly in the field of behavior, and indirectly, through the medium of the dependent effector organs, in the geographical field, precisely because the tension in the field of behavior can only decrease if the distance in the geographical field decreases.[1]

For both gestalists and cyberneticians, the attraction in the behavioral field is, after all, entirely analogous to the physical phenomenon, and it is still a physical connection that unites the two fields and turns physical gestures into a set of dependent systems. *The brain is a kind of machine that enables the principle of least action in a domain in which it would not naturally occur.*

When a baby sees an object and picks it up to put it in its mouth, there is supposed to be only an instantaneous dynamic effect. There is no need to mention a nonactual instinct, a transcendent and protean libido, a "meaning" of the gesture, which cannot be reduced to its physical actuality. The driving force of the gesture is provided by physiological energy reserves, and its direction, by nervous mechanisms or dynamisms functioning through an

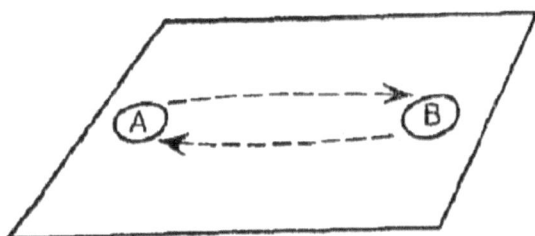

Figure 3.1.

energetic derivation similar to that which operates the steering servomechanism mounted on a more powerful machine.

A careful and faithful presentation of the thesis is enough to reveal its implausibility, and more importantly—which is also more interesting—to reveal the point at which the essential correction must be applied. The behavioral space can only play its role if it is given a hyper-geometric and hyper-physical "dimension." Unlike physical feedback or self-distributions, psychological and axiological feedback and self-regulations can only be conceived if they are immersed in a non-geometric "dimension," the properties of which are irreducible to those of physical space. A "seen object" can only attract the "hand organ" if it has a *meaning*, with regard to either a conscious or unconscious need, or an intention or instinct aimed at a nonactualized goal. On the other hand, and simultaneously, the "distance" in the psychological field, from the "hand image" to the "object image," can only regulate the course of the action through psychological feedback if it is a "seen distance," according to absolute survey. It cannot be a "step-by-step distance," in which the nervous mechanism would be regulated only by the differential dynamic effects, through feedback, just like a radar controlling a D.C.A. cannon. Moreover, the dynamics and kinematics of psychological feedback imply a hyper-dimension, an axiological space, combined with space and physical dimensions. The field of behavior—and possibly its objective equivalent, which appears to us as the cortex—is not simply a kind of "analogical machine," functioning on the principle of least action; a guide table, a servomechanism controlling the mechanics of the body; it is a "converter" between the axiological space and the physical space. What is known as consciousness is the very act of conversion.

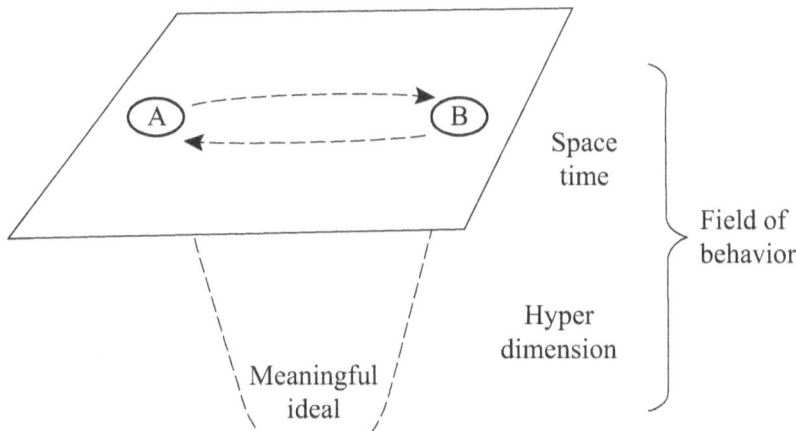

Figure 3.2.

Therefore, we cannot schematize psychological feedback, where meaning and value intervene, on a single spatiotemporal level, even if this level is supposed to represent the four dimensions of space-time. It is essential to include the hyper-dimension, the connection of which with space provides both the dynamism of meaning, controlling the expenditure of energy extracted from physiology, and the kinematic regulation of action by "information" in the nonmetaphorical sense of the word.

There is a contradiction in imagining that cerebral processes still obey a banal, "molar," and blind dynamism, or that the field of consciousness is only a place of attraction or repulsion. I see both my hand A and the object B *at the same time* when I move my hand to the object; I do not only feel a "solid" dynamic release when my hand approaches the object. Moreover, the direction (*Richtung*) of the act, in its aspect of geometrical movement, is enveloped by its meaning,* at least implicitly. Without this "enveloping meaning," its dynamism would be only a "monotonous function" of the distance between the hand and the object, and any detour would be impossible. In the absence of meaning, it would even lose all dynamism. If I no longer know why I make a movement, the movement soon stops, unless a subconscious thematism, still charged with "meaning," relays my conscious intention. Ultimately, there must come into play an absolute survey of the situation, beyond the auxiliary effector mechanisms, along with a combined perception of the geometric situation and the meaning, or the value achievable starting from this geometric situation. In order for it to work, a trans-actual ideal must enter the circuit of the nervous feedback. In other words, the guiding information must be something other than a stimulus. The organic system must be "dependent," not only on its nervous apparatus, but on a trans-spatial ideal of which the cerebral feedbacks are only auxiliaries. Lastly, this trans-spatial ideal must be directly dynamic, no matter how quantitatively small this dynamism may be compared to the subordinate dynamisms that amplify its effects in the physical world.

[THE THEORY OF KURT LEWIN

Unlike cyberneticians and orthodox gestaltists, Kurt Lewin,[2] while striving to depict psycho-organic behaviors using vectorial and topological models, did not claim to interpret these schemas as representing physiological reality, the brain field,* supposedly isomorphous to the field of behavior. For him, the vectorial and topological model directly and exclusively represents psychological reality. He recognized only that the structure of the brain field coincides with the structure of the space of behavior in its essential traits. But Lewin, just like the orthodox gestaltists and the cyberneticians, only insisted more vigorously on the rejection of all psychological explanation, which would appeal to some entities "behind" the field of actual processes, such as instinct or will. The task of dynamic psychology, he writes, is "to find the psychological laws and to represent the situation in such a way that the actual events can be derived from it."³ In that, psychology enables the transition, as the physical sciences did in their time, from the Aristotelean state to the Galilean and Newtonian state. In the epoch of Galileo and Newton, physics changed the meaning of the word "explanation." It rejected theories of the ancient type, which sought explanation not in the relations of dynamic events [*faits*] themselves, but in entities situated behind the events.

Lewin's conception does not essentially differ from that of the cyberneticians and gestaltists. It rests on the same "actualist" hypothesis. Since Lewin states these postulates with clarity and precision, it is interesting to examine them with particular care. This will allow us to better define, *a contrario*, the axiological space. The two postulates that properly represent the life space are as follows:

a. The principle of "*concreteness*"*: "Effects can be produced only by what is concrete, *i.e.*, by something that has the position of an individual fact which exists at a certain moment; a fact which makes up a real part of the life space and which can be given a definite place in the representation of the psychological situation."⁴ This principle excludes all explanations by development, adaptation, *wirkende Secle*, tendency, or instinct in its abstract definition.⁵

b. The principle of "*contemporaneity*." This results from the preceding principle. Neither a past nor a future psychological event can influence present events. Only the actual can act on the actual. It was typical of the Aristotelian way of reasoning not to sufficiently distinguish historical questions from systematic questions (in contrast to the dynamic mode of explanation of Galileo-Newton). As a result, it considered some past or future events as possible causes for present events. Finalist explanations, based on future causes, like historical explanations, based on past causes, violate the principle of contemporaneity. Even the experimental psychologists continue to be Aristoteleans without knowing it. For example, by appealing to an instinct or tendency aiming toward a future state, such as considering children's play as a sort of practice, one implicitly allows an action of the future. Symmetrically, many theories of the expression of emotions, based on phylogenetic identity and not on the similarity of situations, postulate, when it comes down to it, that the past acts. They poorly hide this postulate by appealing to memory as the bridge between the past and the present.

Lewin's error is palpable. In favor of the evidence, that only the present is *actual* [*actuel*] in the two senses of the word—that is, that it takes place now and is acting[6]—we are oblivious to the existence of a "dimension," ranging from the ideal to the actual, from the abstract to the concrete, completely independent of the dimension past-present-future.

The first postulate is in reality drawn from the second, even though Lewin believes precisely the opposite. But the axiological space is beyond the time of the physicists as well as beyond their space. If we represent the space-time of the physicists by this schema (a single plane in perspective designating the four dimensions—see figure 3.3), it is clear that neither the past nor the future can be active. But nothing prevents the actual present from being in dynamic equilibrium with a trans-temporal and trans-spatial ideal that would intervene in the circuit of an axiological feedback. When an animal acts by instinct—when it is in the proper sense "overtaken" by an instinct—it is not necessary to tie this instinct, as Lamarckism believes, to the history of the species, or, as naïve finalism believes, to a divine intention taking into account the future. But it is necessary to take into account something other than the actual present state of the organism, given here and now. The impromptu postulate would, rather, immediately assume the existence of either an unobservable microstructure in the organism (instinctive activity being only the operation of this microstructure), or a field of forces of the same nature as the fields of attraction and repulsion of Newtonian physics.

This is even more apparent if we consider the work of the formative instincts. The principle of *concreteness* is inapplicable, or more precisely, it becomes insignificant, when applied to embryonic development. This development is incomprehensible if we only consider actual functioning or equilibrium. The interpretation according to action of the past (mnemism[7] in the manner of Haeckel) or according to action of the future (naïve finalism) may be false, and to a large extent they probably are. But the purely actualist interpretations—that is, those which appeal only to immediate structural or

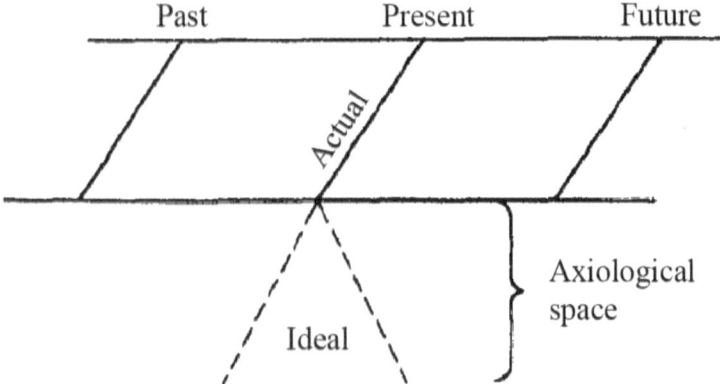

Figure 3.3.

dynamic states, excluding all trans-spatial potential and all organizing dynamism—are certainly entirely false. Because epigenesis is an experimentally demonstrated fact, while preformationism, in all its forms, is a hypothesis that has been eliminated.

Contrary to the principle of concreteness, there is something behind the actual processes. They are in dynamic interaction, not only with each other, but with the formative themes located outside of space-time. Lewin fails to recognize the primary character of the force in the microphysical or psycho-biological individualities. He fails to recognize that the forces of classical physics are merely statistical averages of these primary forces. He gets things wrong by wanting to interpret the dynamism of individual activities according to explanations of the Galileo-Newtonian type. Failing to recognize their primary character, he fails to recognize their true nature. The forces in an individual activity derive directly from the joining end to end of the actual with the trans-spatio-temporal; they express the ideal → actual tension. If the forces of physics seem to express actual → actual tensions, it is quite simply because the individual microphysical actions that constitute them are lost and neutralized in the general statistical effect. Nervous feedback seems to function only in the actual and according to tensions which go from actual to actual, but they are enveloped and framed by an ideal → actual tension. And in the case where they are effectively reduced to actual functionings, their assembly at least has been operated by a primary feedback with a trans-spatial component.]

AXIOLOGICAL RELIEF

In most actions and perceptions, guiding values that are more or less embodied in psychological themes, and meanings transcending space that are more or less embodied in "knowledge," and linked to the here and now, underlie the dynamism that [Lewin believes][8] to be representable in the single plane of the actual. If I see from afar a child playing without supervision at the edge of a high cliff, the more dangerously close he comes to the edge, the more intense the impulse I feel to rush in and hold him back. If there is a guardrail, I don't feel any such impulse, as long as I know it is secure. The dynamic effect of the cliff edge and the inhibiting effect of the image of the guardrail on my reactions are obviously unintelligible if one disregards the meaningful "knowledge" that transfigures their image. These meanings and values do not alter the equilibrium of the elements of current perception, in the way that mechanical memories alter mechanical feedbacks when the automaton's program controls their activation, or the way the attraction of a celestial body is combined with the attraction of other bodies. These meanings and values are a sort of permanent presence behind the spatiotemporal scene, giving it as a result a kind of *axiological relief*, analogous to the impression of depth produced by the combination of two images in stereoscopic vision.

This "axiological relief" manifests the psychological effect of the hyper-geometric dimension. It concerns "the important" or "the valuable" in the same way that ordinary relief concerns depth. It is manifested by organized feelings or emotions, by an impression of "intensity of importance," in the same way that ordinary relief is manifested by an impression of stepped distance. It may give us a feeling of emotional vertigo, similar to the vertigo experienced with depth, when there's a sudden drop in the depth of the important or the valuable, and when we feel that a slight difference in our spatiotemporal behavior will lead to a vital difference in our axiological equilibrium. On the edge of a cliff, a single step in the horizontal direction can result in a deadly vertical fall. Similarly, sometimes a single word, a single gesture, can either lose or save us, in the "dimension" of meanings and values.

Let's take an example from Stendhal that is discussed by Lewin.[9] Julien has decided to marry Ms. de Rénal. However, when the time has come for him to leave, he does not yet have the courage to carry out his plan. At a quarter to ten, in an anguish that almost makes him lose his mind, he says to himself, "When the clock strikes ten, I'll do as I've decided, or else I'll blow my brains out." This example seems to illustrate that a future event can have a powerful influence on behavior. Lewin, however, contests this, asking, is it really the future? If a child is trying to catch a toy that is visible, but hard to reach, the goal is certainly psychologically present. It is obvious that, for Julien, the goal "to marry Ms. de Rénal" is likewise part of his present "life space." It is only the object-content of the present psychological goal that is in the future, as a physical or social fact. The present psychological reality of feelings such as fear, hope, and doubt, does not depend on whether the object-content of such feelings exists in a physical or social sense.

These observations are indisputable. But what, for Julien, is "behind" the present situation, troubling him to the point of anguish and madness, is not the future, it is a meaningful ideal. Julien's resolution is related to his ideal of life; it has meaning only through him. The hero's mental vertigo is a case, if ever there was one, of vertigo in the face of an "axiological relief." With a small move he is either lost or saved, just like that child on the cliff. The "life space" is only what it is because of the "life ideal" that envelops it. Marrying Ms. de Rénal is, in itself, as an image in Julien's consciousness, neither attractive nor repulsive. It is attractive only because marrying her has the meaning, for Julien, of a "heroic conquest." It is repulsive only because it has the meaning of an "inappropriate and dangerous act." The anguishing ambivalence of the projected action is conceivable only because of its connection with two antagonistic meanings, at differing distances in the axiological space, each giving rise, separately, to a particular axiological dynamism. The fact of ambivalence alone can be a decisive objection to "flat" geometric schemas in psychology. In the life space, we will admit, the future does not really intervene, and everything is "actual," that is, present. But the present is dynamic only because it is in tension with a guiding ideal that is trans-actual. Julien's resolution, "to marry Ms. de Rénal at ten o'clock," is only a projection in space and time of this ideal among thousands of others possible. The properties of the projection can only be explained by the properties of total reality, in the same way that the shadows, in a flat drawing of a sphere, only make sense because the real sphere is actually three-dimensional.

UNDETERMINED IDEALS

In the absence of the axiological dimension, it is difficult to understand how it is that most of our goals can be undetermined. We do not see at all how it is possible to include these undetermined goals in a topological "space of behavior," or in the schema of an exclusively nervous feedback. This difficulty is not solved by pointing to the distinction between the indetermination of the *content* of the ideal goal, and the concrete determination of the *psychological* fact itself. Let us take the case of an artist, or an inventor, haunted to the point of anguish by an aesthetic or scientific ideal that he foresees but is unable to achieve. One may say that as a psychological fact, this tension or anguish is nonetheless perfectly determined [and this is enough to safeguard the principle of concreteness]. But this is merely a play on words. No one denies that the embryo, in each of the phases of its development, is something concrete and well determined. But this "concrete" is nonetheless abstract and undetermined compared to the adult stage or to a more advanced phase of differentiation. It is exactly the same in the case of the artist or inventor. His hunches are specific psychological facts; they are what they are. But one must add at once: they are what they are only because they are heading toward an ideal that has not yet been realized. This hyper-geometric transitivity, or, if one wants, hyper-concrete transitivity, is part of their very being, and alone explains their dynamic character. It is not the presentiment of the goal, it is the goal of which there is a presentiment, which attracts the artist. [On a vectorial schema of the space of behavior, it is not the presentiment that it is necessary to represent, it is the ideal aim, and an ideal aim must be symbolically represented on a dimension that is "vertical" to spatiotemporal dimensions, and is of an order entirely different to them. If we want to respect the principle of concreteness, by representing only the presentiment, it is no longer dynamic. If we want to respect the dynamism, it is necessary to abandon the principle of concreteness, by representing the goal outside the field of the actual.

We know that Jean-Antoine Watteau created the composition of *The Embarkation for Cythera* slowly and progressively.[10] He refined his "arabesque" over a long period of time. If he had died more prematurely, without having time to realize this work, the final arabesque of this picture, such as we have it, would not have had any place in our world—and yet it would have acted on our world as an invisible center of attraction behind the painter's various drafts. At no time would it have been "the future," relative to the present [*l'actuel*], since, by hypothesis, it would never have been realized. This is proof that the temporal dimension past → future is only an inadequate projection, in such a case, of the hyper-dimension ideal → actual. An artist is struck by a subject, which will become a work [*oeuvre*], if God gives it life, only because it awakens in him a still vague, but dynamic, ideal.] Mathematicians who, before Lobachevsky, tried to prove the postulate of parallels, were guided by an ideal that was determined only roughly: "To find an unassailable demonstration." Moreover, under the circumstances, this roughly expressed ideal was not the authentic ideal, which was "Non-Euclidean geometry." Far from being able to be schematized, the ideal only had a deceptive effect in the researchers' psychological reality. And yet, it is this kind of unobservable ideal that governs the lives of countless artists or inventors. It is a moral or social ideal that is unobservable, or hidden under utopian disguise, that drives the lives of countless activists and keeps them fighting against all odds. Winds and tides, that is, actual forces or obstacles, can be represented without difficulty in the space of behavior, but the ideal that counterbalances them cannot, because it cannot be located anywhere (without, however, being a utopian Nowhere*).

Regardless of the angle chosen for its study, the problem of psycho-biological action is unsolvable unless the hyper-geometric dimension is taken into account. Let us take, for example, the idea of possibility. "The fundamental constructs which we use in representing the situation must consist of concepts from which one can derive unambiguously, certain events as "possible," others as "not possible."[11] But the man who seeks an ideal does not always know whether or not it can be realized. If it is only a matter of taking a full glass on the table without knocking it over, in the visible absence of obstacles, it is as if the possibility is inscribed in the field of vision. When it comes down to it, this possibility is already trans-physical, as it only defines itself by "survey" and "knowledge," but ultimately it is easily calculable. The success of the feedback is unambiguously predictable. But an inventor, by definition, can never clearly know, in his "life space," whether what he is looking for is possible. The moment he perceives this possibility, his problem, by definition, is virtually solved. Possibility is not a simple concept. It has stages that correspond to different "distances" along the hyper-geometric dimension. If the possibilities were included in the field, and if the dynamic situation was fully representable, the activity would stop as soon as the equilibrium position is reached. The "actualists" are indeed compelled to recognize that this consequence is contradicted by the facts, especially in the case of human beings. David Krech and Richard S. Crutchfield, while affirming—with Lewin, whom they claim to follow—that "there is a constant tendency for the psychological field to change in the direction of reduction of tension," are indeed compelled to add that this *"does not imply that the achievement of a state of equilibrium is the goal of the individual's action."*[12] But it is difficult to see how this incontestable fact can be reconciled with their hypothesis.

[Krech and Crutchfield attempt two explanations, each very different. The first is mechanist. While in simple physico-chemical systems the final state of equilibrium is the same as the initial state—for example, when the blood's pH has unbalanced due to intense activity, the respiration speeds up in order to return the pH to its initial level—in the psychological field, the final equilibrium is almost always different to the initial equilibrium. Its history is not that of a single fluctuating system, always returning to the same position. "It is, rather, a history of changing equilibria, in which the psychological field restructures continuously."[13] This explanation is clearly insufficient. It does not answer the question, which is to know why the achievement of an equilibrium state is not the goal of an individual's activity. Assuming that the activity is controlled by a set of multiple, interrelated feedback loops, like Ashby's homeostat, it would still be seeking successive equilibria. And it would not be its own fault, but that of the environment [*milieu*], if a final equilibrium were never found.

The second explanation is very different. Krech and Crutchfield appeal to the person's progressive physical and psychological differentiation, and his efforts to integrate increasingly complex states through a hierarchical and mobile system of values. The individual's "level of aspiration" is raised as he achieves his first goals. He "continuously sets for himself new levels of accomplishment (his levels of aspiration) that are above those of his present achievements."[14] This is absolutely true, but it amounts to the complete rejection of the initial, actualist hypothesis. Krech and Crutchfield can't help but notice: "[T]his phenomenon seems to contradict a simple theory that behaviour tends toward equilibrium; here it seems that an equilibrium is voluntarily abandoned and a tension deliberately set up. The person seems to be pulling himself up by his own bootstraps."[15] Clearly, the sole conclusion possible here is to reject Lewin's fallacious postulates, and the bias of a spatiotemporal dynamism.]

IDEAL BARRIERS

The barriers that prevent the student from solving a mathematics problem, although he is "moving" toward the solution, cannot be put on the same level as physical barriers—or even social barriers, although these always have an ideal component—which prevent him from going out to play. There are barriers between the space-time world and the hyper-dimensional world. Clearly, humanity has to overcome some barriers in order to conquer the still-uncharted world of future technics. The discoveries of electronics, plastics, nuclear energy, or plutonium were not faced with the same kind of challenges as those that prevented Europeans, before Columbus, from reaching America. We also have, in the present state of technics, more or less difficult access to trans-space, to free regions, and more or less strong barriers, with some communications and pathways. Biological evolution has already taken place through successive conquests and the overcoming of barriers in the trans-spatial domain. Over the last few centuries, it has been achieved much more through technical conquests, in the fashion of Western science, than through geographical conquests in the fashion of Columbus's discovery of America. It is almost always the annexation of a technical domain which then allows geographical annexation, in both the biological and the cultural evolution of mankind. Ideal locomotion precedes and enables physical locomotion. Marine organisms only invaded land after learning to breathe oxygen from the air. Man only moved through the atmosphere after having found the combustion engine, and the jet engine opened up the possibility of interplanetary travel. Cybernetics itself represents a field of possibilities that had to be conquered by this ideal locomotion in the trans-spatial that is invention. The "cephalization" of human industry continues the cephalization of vertebrates and simians, and it ensures the control of mankind over the geographical environment. The very success of cybernetics conceals the condition of this success, for the same general reasons that industrial technology, the triumph of the inventive spirit, and the conquering of the trans-spatial, appear to superficial minds as the triumph of materialism. It is this same general illusion that makes mankind believe that intelligence and reason are autonomous, completely cut off from animal instinct, which is nevertheless the primary manifestation of the connection between space-time and the trans-spatial.

Chapter 4

Communication

Communication between two people, *A* and *B*—that is, the transmission of information from one to the other—operates according to a schema completely analogous to that of the action of an individual, *A*, aiming at an end, and using a subordinate function. If *A* wants to communicate to *B* an idea that he just had, he expresses it with a phrase, if he is neither mute nor aphasic, a phrase that is manifest physically in modulated airwaves, and in a series of electrical waves if he uses the telephone. These series of waves are retransformed into meaningful themes by *B*, if he is neither deaf nor agnosic, and finally these meaningful themes again become *A*'s idea, in *B*.

As it is a matter of two conscious centers, and not of a conscious center and an ideal, the symmetry between 1 and 4, 1a and 3a, 2 and 3, is even more perfect than in the schema of a simple activity, and the whole is easily reversible. Communication machines are mechanically reversible: the telephone, the three-electrode lamp, and the apparatus of phonographic recording can emit as well as receive signals, at the price of minor modifications. The biophysiological apparatus of the head of a human being is equally reversible,

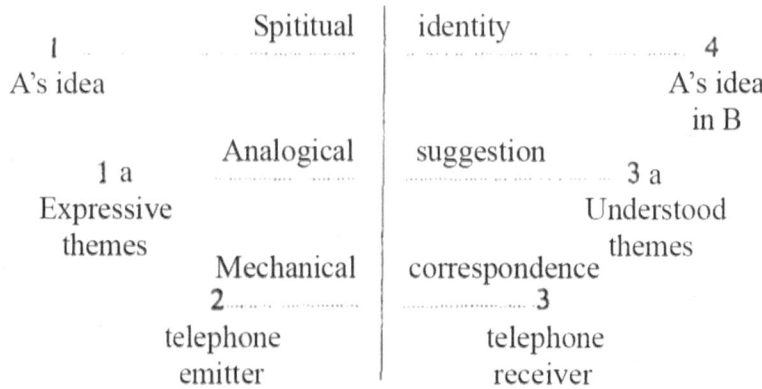

Figure 4.1.

according to a mode that it is difficult to conceive as mechanical: it can transform an idea into expressive themes and words in order to express itself, and it can transform words or themes into ideas in order to understand. The route of understanding is not exactly the route of expression. It passes by the ear and the sensory zone, not by the larynx and the motor zone. But the "meaningful themes" and the "understood themes," as psychic realities, are isomorphic, and they sometimes model themselves directly on each other by analogical suggestion.

As to the finally understood idea, it is not only analogically "the same" as the expressed idea, it is absolutely the same idea, which is spiritual, save for psychic perturbations. And it would be absurd to speak of mechanical reversibility in its subject. The individuation of the idea in the psyches of A and B must not mislead us: it remains, essentially, a single and same idea. If the spiritual "I" of A and B, in opposition to their psychic "me" are not, like the idea, absolutely "one," they at least tend toward unity, and they would doubtless attain it if they were purely spiritual. This at least is the dream of all mystics, beyond all technique. And the mystic is right because without the ideal identity of the "I" and ideas, no communication technique would be possible, just as, without "temporal survey" and the relative eternity of the "I," no technique of action would be conceivable.

In the case of communication, as of action, the axis of symmetry for the entire system is the present [*actuel*] functioning, here and now, of a framed machine. If the speech acts of A are recorded on disc, a time of variable duration can insert itself in this "present." The structural inertia of the disc, probably like inertia in general, certainly implies a very complex (not very simple, as we have long believed) relation to space-time. But this complexity changes nothing fundamental in the phenomenon.

There exists an intermediary case between individual action and inter-individual communication: that of mnemic self-consultation. If I have an idea, I "entrust it" to my memory, and sometimes I make some summary notes to help me retrieve it. Next, I consult myself, with the help of the notes if needed, and if all goes well, I retrieve the "same idea." The inertia of the manuscript notes, or mechanical or magnetic recordings, and the maintenance of some psychic sets that constitute the psycho-biological part of the memory—which is certainly of a nature entirely different to mechanical inertia—make a bridge between the idea first invented and the retrieved idea, between "I" and "I."

Naturally, positivist or mechanist psychology, particularly psychology inspired by cybernetics, treats the problem of communication exactly like it treats the problem of action. It exclusively considers the intermediary, framed part of communication, and claims that the framing part can be treated as being of the same nature. In the same way that it reduces action to the functioning of feedback, it reduces communication between consciousnesses to a structured transmission, going from an element A, the transmitter, to an element B, the receiver. At the limit, one can consider material transport as a form of communication, even though in general communication is a transport of information, not of material. According to cybernetics, there is no more reason to make consciousness enter into the theory of communication than into the theory of information. Communication does not necessarily take place from person to person. It can be from machine to machine, or from one part of a machine to another part. Suppose that a meteorological station, A, records hourly the temperature, barometric pressure, and wind speed, and that a caretaker telephones the recorded data to a station B. If the situation of the caretaker in station A becomes too uncomfortable, we can very well replace him by an automaton who will communicate via electrical transmission with station B. Suppose that this station is tasked with using the information received to guide airplanes via radio. Once again, an automaton could replace personnel there. The airplanes themselves will one day be able to fly without a pilot. If we speak of information and communication when there are workers at the stations, or pilots on the airplanes, we do not see why it would become illegitimate to do so again when the workers have been replaced by automata. What is advantageously replaceable in a system cannot be the essential part of the system.

This thesis is as unacceptable in the case of communication as it is in the case of action. It is quite evident that the only irreplaceables are, on the contrary, the subjects who act, exchange messages, and oversee their various technologies. A pure transmission of informing structure[1] can only become information and communication when the "support"[2] is the expression of a meaning conceived by a consciousness and is the occasion of a grasp of meaning by another consciousness. The spatiotemporal part of informing communication can be reduced almost to the point of evanescence. Two spiritual consciousnesses, nearly identical, even granted a psychical "I" and "you," understand each other almost without exchange of speech or signs. By an exaggeration which is the reverse of that of the mechanists—which does not prevent their descriptions of the communications of consciousness being very superior to those of the cyberneticians—the phenomenologists and existentialists (Husserl, Scheler, Hartmann, Marcel, Buber, Nédoncelle), like the mystics, go to the extreme and believe that we can do away with all intermediary material, even psychic, in the communication of consciousnesses.

The truth is that the intention of communicating, from an "I" to a "you," is in effect more essential than the communication technique. The innumerable and quite vain theories on the origin of language have in general accorded a lot more importance to the presumed occasions which were able to give birth to such and such a technique of expression. The truth is that language exists virtually from the very instant when two people, face to face, become conscious of being a personal "you" and "I" (or better, according to the ingenious theory of G.H. Mead, of being interchangeable "you's"). Man himself is quickly set up to speak as soon as he can want "to signify." And he can want "to signify" as soon as he has seen, in one of his fellows, a conscious being.

Incidentally, the inter-communications of two animals also exceeds the purely physical transmission of signals. They remain in the zone of psycho-biology and instinct, they suppose instinctive knowledges and practices rather than consciousnesses grasping meanings, but they also suppose a passage in the trans-spatial world, and cannot be reduced to the mechanical "median part."

[Let us represent reality as a whole by the surface of a sphere, and the conventional space-time on which classical physics claims to project all real phenomena, even those which are beyond space, by a cylinder tangent to the equator of the sphere (figure 4.2). Contrary to the theses of the mystics and of certain phenomenologists, all informing communication necessarily touches the spatiotemporal equator through its median part, even if this is small. But a part of the "trajectory" of the communication necessarily passes through the "meridian" zones of the sphere and is only artificially projectable onto the cylinder.

In the same way that an automaton cannot truly reason, but only mechanically carry out the "combinatory" of reasoning, it cannot truly transmit

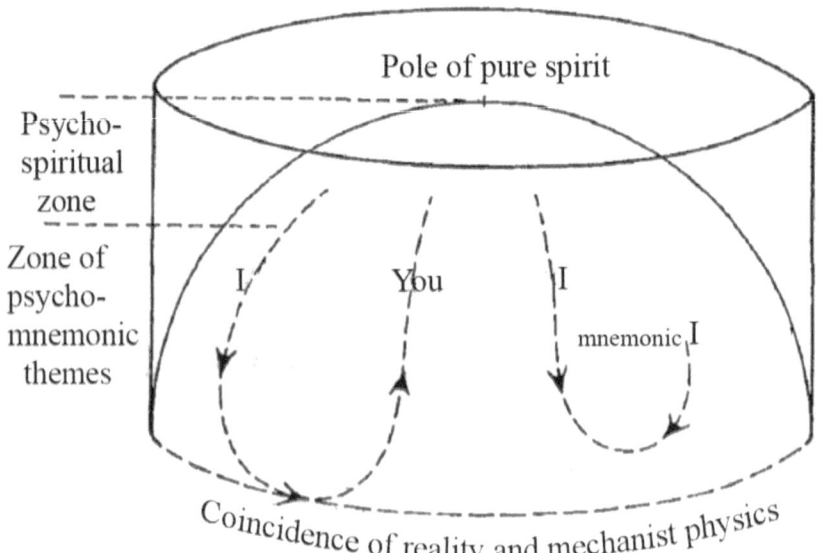

Figure 4.2.

information. It cannot truly communicate but only carry out the mechanical part of the transmission. But beyond the combinatory, there is the "noegenetic,"[3] which envelops it. Beyond the transmission of the pattern, there is the intention "to signify," the will to modify a consciousness. Communication is always "persuasive"; it always has something, if not rhetorical, then at least axiological about it.]

Already all perception, whether human or animal, is a mixture [*mixte*]. In its physiological origins it involves an aspect of purely mechanical or physical communication, which can probably always be reduced to a certain order[4] of impacts of photons. On the other hand, it is consciousness, or, if one prefers, the apprehension of meaning or of beings in their signification.

If instinctive or intelligent perception is already impossible to represent entirely on the equatorial line of the schema,[5] this is even more so with the perception and comprehension of a language—because all authentic communication is a language. A language always implies an ensemble of mechanical and physiological media of communication, functioning in the spatiotemporal plane [tangent to the equator], a "horizontal" trajectory. On the other hand it implies two conscious centers, the emitter and the receiver, capable of expression and of comprehension: that is to say, capable of "vertical," trans-physical participation with a world of ideas, and capable of converting the ideas into structures[6] and the structures into ideas. It also implies a code, more or less instantiated in habits or memories, psychological pathways or conventional channels, guiding the vertical participation and facilitating the invention inherent to expression or understanding.

Cybernetics, which denies this vertical, trans-physical dimension, also of course denies the specifically psychological character of memory and codes. It believes that all memory, considered as a simple stockage of structures,[7] can be imitated by mechanical models, and that it does not have to be interpreted as dealing with some meaning. It believes that the mechanical models of memory permit us precisely to reject the very idea of a trans-physical domain of meaning as a mystical and useless hypothesis. Just as it considers the mechanical communication of structures not as auxiliary, but as the entirety of communication, it considers psychological memory not as the auxiliary of the vertical circulation along the trans-physical dimension that we have defined, but as the whole of what, in information, appears to be added by the receiver to communicated structures. The meaning of information is the functioning that it controls. The meaning of a communication according to a code is the ensemble of stored structures[8] that it "unlocks" ["*déchroche*"]. Cybernetics confuses a mere unlocking of materially recorded and stored information with the reminder of a mnemonic potential, linked to a meaning, that cannot exist in space.

Electronic calculating machines have "memories" capable of storing a great amount of information: systems of electric valves opening and closing, magnetic tapes, mercury tubes, and so on. These mechanical recorders are supposed to be true models of psycho-biological memory. And the methods used to activate these memories, which depend on the machine's own assembly, are supposed to be close to our codes. Thus the "mechanical memory" machines would be capable of correcting printing or spelling mistakes, on the condition that they have a sufficient store of models of correctly written words, and also that the words to be corrected are not too badly spelled.

The confusion of an unlocking, an activation, with a mnemonic reminder is clumsy. The mnemonic reminders provoked by the reception of a message have nothing in common with the utilization of information stored by a machine: they are indissociable from the recreation of a meaning, according to the intellectual capacity of the receiver—that is, according to whether he is connected with a world more or less rich in ideas and meanings, not more or less well-stocked with "typographical cases."

WELLS'S UTOPIA

In his utopian novel *Men Like Gods*, H. G. Wells supposes that some English and French Earthlings, having arrived on an unknown planet, listen to a speech by one of its inhabitants, who first explains to them what his function is. To their great surprise, each of the Earthlings can understand as if the Utopian speaks their native language. But each has understood according to his culture and the level of his intelligence. One has understood "I study the action of nuclear fields on electrons," and another, "I weigh solid bodies." This utopia of Wells represents a philosophy of language and information that is better than Norbert Wiener's theory. And it is not very far from the facts. Language between people is often similar to a biological "induction": the same chemical substance determines some very dissimilar differentiations, according to the embryonic areas or the tissues affected. It is similar to a sort of very general invitation to understand. We always try to find a meaning in the messages we receive when they appear to us to be obscure or contradictory. Everyone who is informed always informs themselves by themselves, through an irreducible invention.

Linguistic codes never proceed from an automatic correspondence with their pure state, and the invention of codes would be inconceivable if a certain primary information could not operate in an almost immediate spiritual communion, without code, and with a minimum of spatial transmission. In the same way, the mechanical auxiliaries of memory would be inconceivable if memory were purely mechanical. When, as we say, we consult our own memory, the current "I" is not like a person who goes to a closed register in his library to open and decipher it. The "I" consults itself, without an interposed means of communication; or, more probably, with some entirely psychological means. [It is as if the line of communication, without descending this time to the equator, nevertheless passes by a more equatorial zone.] The "I" does not have to understand how to read itself. Its memory supplies it almost directly with a meaning, conveyed sometimes by some auxiliary psychological themes, but independently of all code and all deciphering.

Between the current "I" and the mnemonic "I," there is no dialogue. Because, even when the mnemonic effort is accompanied by interior language, it is not the mnemonic "I" that speaks like an "other," which we hear. It is the current "I" that interrogates itself in order to aid the "telepathic" effort of reunion of the current "I" and the mnemonic "I," which were provisionally disjoined. In the mnemonic effort, like in the effort of invention—where the "I" participates in a sort of "universal I"—the interior dialogue is certainly not essential, and is often in fact absent, as experiments have demonstrated.

Yet communication and information between two people, or even between two psycho-biological individuals, is not as different to mnemonic auto-consultation as we would believe. Two living individuals, participating in the same specific memory, are not two absolutely distinct individuals—otherwise, crossed reproduction or double heredity (paternal and maternal) would not be possible. Most importantly, two people participate in the same world of ideas, which is familiar to both, dialoguing a little like a man deliberating with himself or consulting his own memory. They are a little like a mnemonic "I" for each other. The interposed mechanisms barely count.

WIENER'S MYTH

However, just as there are momentary self-alienations in the conscious being, when he is seized by a mechanism he no longer controls, could there not be an alienation or inversion of the same type in communication? That is what Norbert Wiener claims in a curious predictive myth.[9] In place of two individuals, telephoning messages to each other across the Atlantic, one can imagine—since material transport is only a particular case of communication—that one individual "telephones" himself with a machine. Thanks to communications via radio-television and Ultrafax, an architect who is in Europe can very well supervise the construction of a house in America, by sending his instructions and plans to a constructor, who reports back to him at every moment on the state of the work. He thus acts elsewhere than where he is materially. In this operation, the constructor plays the indispensable role of being a receiving centre. Through communication, the architect transports his own ideas into the head of the constructor. Up to the present time he needs this other head, which is conscious. But if everything is mechanical in communication, including the two "end points," can't we conceive the possibility of a transport by communication, no longer only of the architect's instructions or plans but of the architect himself? We send our voice by telephone, so why couldn't we send our larynx by a sophisticated telephone? In the mechanist hypothesis, the larynx, and the entire organism, are only a pattern, like the voice. Instead of taking a boat or a plane, the architect would—if we can put it this way—telegraph himself to America. An automatic reader would decode his organism in Europe, destroying him in the process, and a receiving apparatus, just as automatic, would make him appear in the New World. Thus two consciousnesses would no longer be indispensable as contact points in communication. There would no longer be communicators; there would be pure communication. Wiener's myth is exactly the reverse of the mystical ideal. It is reverse and symmetrical: the duality of the communicators is resorbed in the mechanical, instead of being resorbed in the mind [*l'esprit*].

This myth is less absurd than it appears at first sight. Contemporary physics, particularly since the era of wave mechanics, effectively tends to efface the differences between material transport and communication. The majority of modern communication is done with waves, since the particles called material as well as the corpuscles of light, electrons as well as photons, are inseparable from series of waves. The myth of information-transport has perhaps, therefore, like the alchemists' dream of making gold, already become reality on the microscopic scale. It is highly doubtful, however, that Louis de Broglie shows much enthusiasm for this fantastic "consequence" of wave mechanics.

Without going as far as Wiener's myth, it is necessary to recognize that in certain cases, which are quite ordinary and common, a semi-alienation is produced in communication or transport. Instead of framing the subordinate transmission mechanisms with his consciousness, he can transform himself into an unconscious parcel. When I travel by sleeper train and I am asleep, I am hardly anything else. It is true that it was my own will that decided my transport, before falling sleeping during the journey; in a way I receive myself in the station of arrival. An analogous but more striking case is that of being anaesthetized for a surgical operation. The surgeon naturally asks the consent of the patient before the operation, but then the patient is treated like an inert corpse. If the patient has a cardiac arrest, and if the surgeon only resuscitates him by massage, he recovers himself on waking a little like he finds himself on the other side of the Atlantic after having traveled across it in wave form. In these latter cases, in reality the principle of framing is not violated. Consciousness frames the voluntary suppression of consciousness during the spatial or temporal transport of the organism.

Chapter 5

The Origin of Information

[R. A. Fischer,] Claude Shannon, and Norbert Wiener were the first to define the notion of the quantity of information with precision.[1] As we have seen, basic information is the alternative "Yes-No," or any other "binary" decision: 1 or 0, right or left, and so forth. This is the unit of information, or "hartley":

Figure 5.1.

If we have to localize a point on a line, or describe a given figure with perfect precision, the quantity of information must be infinite. Practically, a description, a localization, or a measure are never perfect. Let's consider a point P, which we simply know is situated on some part of a line between A and B. We are then informed, by 0 signifying the left half, or by 1 signifying the right half, that it is in the left half, then, in the right half of this left half, and so on. The binary number, in the form 0, 010101 . . . , which expresses our information can never be an indefinite series of 1s and 0s. Information has a limited precision. It only gives a zone of probability ab, between A and B. Here, for example, the fifth digit after the comma is undetermined. The quantity of information is limited to four digits after the comma. The quantity of information gained by the passage from AB to ab amounts to a logarithm of a probability. The formula that expresses it is exactly the formula for entropy, which is also a logarithm of a probability but with the opposite sign. Information is a negative entropy.

This result was surprising and even seemed "sensational." It is, however, only natural and is easy to understand. Suppose that the line AB represents a metal bar at a certain uniform temperature. We raise the temperature of zone *ab*. The second law of thermodynamics implies that the temperatures soon equalize themselves and become uniform again: the entropy of the system increases. While in the previous example, the information about the location of point *P* between *A* and *B* increased as the zone *ab* was narrowed, in the case of the heated bar, on the contrary, the entropy increases as the temperatures equalize and the zone *ab* merges with the whole *AB*. The increase of entropy is then equivalent to a decrease of information, and vice versa. Once again, there is nothing extraordinary in this, since information is synonymous with structure or organization, and entropy synonymous with disorganization. As long as the zone *ab* is at a higher temperature than the rest of the bar, the bar as a whole has a certain degree of structure; it is "informed" in the etymological sense of the word, and therefore, as an observer, I am informed—in the ordinary sense of the word—that a particular event has occurred in *ab*. When entropy has become maximal, the thermal agitation is homogenous throughout the bar, and information is minimal; the zones of probability, previously distinct, are fused. If I write with bad chalk on a bad slate, the rapid homogenization of the chalk dust over the entire surface of the slate is a phenomenon analogous to an increase in entropy and at the same time a loss of information: the written words become illegible. If the "static" on the telephone or the radio reaches a certain intensity, or if the modulations are too weak and descend to a level of fluctuation that causes static, the speech becomes indistinct.

That being said, since any machine, no matter how sophisticated—including calculating machines or feedback automata—can only increase entropy, as they operate according to the principles of thermodynamics, it is evident that they can only decrease information.

The background noise, in the telephone or the radio, can scramble the message, but it is impossible to imagine that the pure chance of the fluctuations can reconstitute a message that has been scrambled or create information out of nothing. It is as impossible, or as improbable, as a kettle freezing when put on to boil. Entropy goes in the direction of the most probable states; information, which has the opposite sign, is therefore an "anti-probability" or, to use an old expression of Arthur Eddington's, an "anti-chance." Chance cannot account for anti-chance. The mechanical communication of information by a machine cannot account for information itself, since the machine can only degrade it, or, at best, preserve it. Cybernetics cannot escape the contradiction. If "no operation by a machine on a message can gain information," and if, on the other hand, "there is no reason . . . why the essential mode of functioning of the living organism should not be the same as that of the automaton,"[2] then where does information come from?

A similar question was raised at the end of the nineteenth century, precisely when the principle of the degradation of energy was popularized. Where, it was wondered, do "higher" forms of energy come from? Immediately, metaphysical and theological answers were drawn, supposed to be valid for the entire universe. They even provided a new proof of the existence of God. Since the universe is like a clock whose spring is unwinding, someone must have wound up the clock in the beginning. The unfashionable "clockmaker God" of the eighteenth century returned in the guise of the "clock winder." God would no longer be the Watchmaker* of Voltaire and Paley, he would be the Winder up.*

But problems that concern the universe as a whole easily arouse the suspicion of being poorly posed, and probably rightly so. There are very few philosophers today, and even fewer scientists, who are much concerned about the "initial winding up" of energy.

The problem of the origin and creation of information is, on the contrary, precise, limited, and pressing. If A speaks to B on the telephone, or A leaves B a message on a slate, the origin of information apparently has some relation with the organism A. The message sent is obviously not created *ex nihilo*. Its conscious sender has himself been informed, instructed, and educated in a social milieu where innumerable instruments of information exist. But in this particular case, he nonetheless plays the role of a source of invention or creation. The one who dictates a message invents more than the one who writes down the message. He is a "winder up" of entropy, to the exact extent that he is an active informant. It is absurd to suppose that only pure transmitters exist in the world. Before communicated information, there is created information. One is almost tempted to stop the discussion here, and to condemn cybernetics without further examination, as one condemns perpetual motion projects without consideration.

However, the cyberneticians and the mechanists have attempted to save their thesis by using the notions of autocatalysis, fluctuation, and coupled systems. Let us examine their "solution."

They first remark—and on this point they are right, in our view—that, if the mathematical formula for the quantity of information and that for entropy are the same, with opposite signs, a dissymmetry nevertheless appears between information and entropy: a machine cannot increase the precision of the information that it transmits, but it can "extend" it. The printing presses of a newspaper cannot correct the typos in the headlines, but they print the article in thousands of copies.

If we consider no longer the quantity, but the extension of information, there is no incompatibility between the (normal) increase of entropy and the extension of information. The printing presses of a newspaper, when they operate and print thousands of copies, consume electrical energy, which they usually degrade into heat. The result of their operation is therefore both the extension of information and the increase of entropy. The chemical phenomenon of autocatalysis, so important in organic structures, must be an extension of information of this kind: the molecule of the autocatalyst imprints, in a way, its form on billions of neighboring molecules.

On the other hand, it is in no way contrary to the principles of thermodynamics that, locally, in a non-isolated system and through energetic exchange with the rest of the world or a bigger system, entropy can be decreased rather than increase, provided that the total increase be higher than the decrease. Well, living beings are clearly non-isolated systems. To say that they can contravene Carnot's principle by implying that they thus have a completely special and extraordinary property is, as the cyberneticians and mechanists emphasize, to play on words. They always represent coupled systems. What would be extraordinary if they were isolated is not at all extraordinary if we consider the "organism-food" or "organism-sunlight" system as a whole. The freezing of water in a freezer is not surprising, as would be the highly improbable freezing of water on a hot stove, because the freezer uses electricity, like the printing presses of a newspaper. Like all heat engines, the freezer creates a local "order" at the price of a larger and more general "disorder."

The photosynthesis of plants is a reaction that miraculously appears to go uphill, instead of downhill. But it should not be forgotten that the plant molecules capture quanta of light that directly provide them with the energy necessary for this uphill climb.[3] The "plant-light" system goes downhill, although the plant considered in isolation goes uphill and appears to increase the amount of free energy by decreasing entropy. Consider a staircase leading to a floor made of fine gravel. If there is no wind or rain, the gravel cannot go back up the stairs. However, a heavy rain can cause many small stones to jump down to the lower steps, and a very small number can jump up by chance to an extraordinarily improbable height.

These notions of coupled systems, fluctuations, and autocatalysis would thus provide a means of understanding how living organisms, while obeying the ordinary laws of physics, can not only extend information but also increase its quantity by increasing the complexity of their structures.[4] One can indeed consider the genes as autocatalytic molecules that imprint their "order" throughout the visible organism. For example, the molecules of chlorophyl can have their "mole" or their "type" in the germ of the plant. And on the other hand, which is the most important thing, the structures of molecule-types themselves can result from an accumulation of mutations provoked by ultraviolet rays, gamma rays, and cosmic rays, which gradually "build them up."

In sum, the phenomenon of photosynthesis provides the principal elements of this attempt to reconcile the irreconcilable. The construction of living organisms is a sort of giant photosynthesis. Organic structures are built up by the accumulation of mutations caused by the more energetic photons and, on the other hand, the flow of energy with which the organisms are "coupled" easily explains the local uphill movements in the operation of these structures. It's as if the water would first form the mill and would then make the wheel turn. Even psychological invention can be conceived as a kind of Brownian motion, concerning not movements of molecules, but cerebral states, some fortunate fluctuations of which are captured and maintained.[5] The man who telephones another is not then, in reality, a creator of information. He is like a mill on a "current of order" or "negative entropy," which he degrades, like any other real system, but by channeling the current in such a way that local uphill movements are temporarily possible. The energy necessary to telephone is taken from food ingested in the previous days and the structure of the message itself. That is, the complex channelings that transform the food energy into "discourse energy" result from organic and social structures constructed progressively since the living species became civilized man, by mutations, selections, and retentions.

[Such is, in sum, the solution proposed by a multitude of contemporary scientists: Erwin Schrodinger, Harold F. Blum, Pierre Auger, Norbert Wiener, and Joseph Needham. It is not essentially different to the old theory of Herbert Spencer, who also "explained" the evolution, that is to say the appearance, of structures, by coupling an *integration of matter* and a *concomitant dissipation of motion*. This "solution" rests on multiple errors and confusions. The only valid thesis that it contains—namely, that entropy and information, despite the symmetry of their mathematical formulae, are not perfectly symmetrical—is precisely what destroys all the rest of the argumentation.

I. The word *order* is equivocal. It designates either a homogenous order (for example, all the water molecules in the current which make the mill wheel turn in the same direction), or a complex structural order in an organization structured at multiple levels (for example, the mill and the body of the miller have an organized structure). The normal evolution toward maximal entropy or disorder destroys order in both senses of the word. In the lower reservoir of the mill, the water molecules move randomly in all directions, and, on the other hand, if nothing contradicts the laws of classical physics, the mill and the miller are destined to lose their structure through ruin and death. Information, on the contrary, is only order in the second sense of the word, despite certain apparent exceptions. Case I: Balzac composed an entire chapter of his *Physiology of Marriage* using randomly chosen printing letters. Case II: Suppose that another author, with a similarly humorous intention, composed an entire chapter using only the letter x. Case III: The reader will have no more information in the one case than in the other, despite the homogeneity of the composition in the second case, and despite the perfect order of the pages, entirely composed of *x*. In order for the printed characters to convey some information, they must be organized into words and phrases, themselves structured by the unity of a meaning.[6]

If we measure the quantity of information in the three cases, we find that the information is equal to 0 in the first case (the Balzac chapter), since the letters were chosen randomly, but that they have the same positive value in the second and third cases. This is contrary to reality, however, since the chapter composed of *x*'s tells us nothing more than Balzac's chapter. More exactly, the quantity of "hartleys" that we attribute to the chapter composed of *x*'s will depend on our conventions. If we consider that each *x* was chosen as a significant letter of the chapter, the quantity of information is the same in both cases. If we assume that the author wanted to compose a chapter using the same letter for the sake of humor, the amount of information should only be measured by the choice of that single letter.

There is no need to reject the mathematical definition of the quantity of information for this reason, but it is important to recognize its superficial and relative nature. In the industrial printing of a European language, we employ around 80 characters. As six binary symbols (1 and 0) have 64 possible arrangements, and seven have 128, each character in a printed book basically corresponds to seven units of information.[7] In order to measure the quantity of information given by a print, it suffices then to multiply by 7 the number of characters used. This way of measuring does not distinguish between Newton's *Principia*, a collection of Careme's sermons, or a history book about Marseilles, any more than a meter distinguishes between silk and cotton. The measure of information, like all measurement, must be made intelligently and relative to a certain mental context. If, for a reedition of Balzac's book, a typographer is obliged to reproduce the typographical "pudding" of the first edition exactly, we can attribute a positive information value to his work relative to this intention. But the typographer would be better not to give himself this difficulty, which is absurd in the circumstances.

We see therefore that, in the second case, the idea of information is applied arbitrarily, and that it is absolutely relative, although we can always measure it. In the third case, on the contrary, information is an absolute reality: the letters form phrases that have a meaning. It is not necessary to object here that, for a Chinese person ignorant of French, there would not be any difference between case I and case III, because we can translate the sense of III, and not the sense of I. It is no more necessary to object that it is again relative at least to the presence of an intelligent reader, and that, in the absence of a person, text III has no more meaning or "absolute information" than II or I—that would return us to the assumption that the absolute of information actually appears not in the person but in the system "text-person." If we again object that text III has only been produced by chance among other possible combinations, and that in the absence of other consciousnesses, there would be no more meaning than with any other fortuitous combinations, we are perfectly right. But with this we then admit, contrary to the thesis itself, that chance on its own cannot create authentic information.

II. We can easily see the importance of this distinction between the two senses of the word *order* for the problem of the origin of information. The supposed initial state of the universe, with free energy maximized and entropy minimized, represents a homogenous order and not a structured order. This initial state resembles text II. It does not contain more information than a pure chaos, which would be similar to text I. We therefore understand and agree with the scientists who have not taken seriously the theologians and philosophers who have claimed to find the trace of the finger of a "clock winder" God in this initial state. We can easily imagine some models of the universe as a curve or a torsion such that the apparent loss of homogenous order finds itself compensated, because homogenous order is always relative to the chosen reference. To give a basic, but clear example, if an explosion takes place at point 0 in a spherical space, the debris of the explosion ends up joining back with itself and converging at the antipodean point, like in the reversed film that represents this explosion. In such a space, entropy, or a related quantity, therefore, only increases up to a certain maximum, after which it decreases. In the same way, the water that makes the mill work can be used again if a lower level is available. It is true that on our planet the sea level marks an impassable limit; but in the universe in general, it is not easy see what could play the role of a "sea level."]

The physicists were therefore right not to be concerned about the theological problem of the entropic upward movement and the origin of homogeneous order. But they would be very wrong to imagine that the problem of the origin of information and structural order is just as untroubling. No curvature or torsion of the universe can automatically rebuild the mill or resurrect the miller.

[III. The profound difference between homogenous order and structured order makes the proposed solution of "coupled systems" completely futile. For it to be effective, the coupled elements must be of the same nature.]A man whose job is to write letters may be assisted in his work by a secretary, or by a guide to commercial correspondence, but he is not sensibly aided by increased nutrition and a flow of "free energy" that passes into his system. This current of "negative entropy" does not help him put his ideas and phrases in order—it may even do the opposite, if he eats too much and has digestive problems. Of course, if he is deprived of nourishment, if he rations his food and does not get enough calories, he cannot work, just as a mill stops if the current dries up. But we must not confuse the dynamic and the kinematic, the flow of force and structural development. Stopping the current stops the mill but does not make it fall to ruin. The return of the current starts it again but does not repair it if it is damaged. The living organism, being mostly constituted of subordinated machines, obeys the general laws of energetics. If I have a horse that bears a load, it is necessary that I feed it, just as it is necessary that I put fuel in a car, if I want it to climb a hill. A horse, and even a person who climbs a hill, only superficially contravenes the second law of thermodynamics, since it is "coupled" with the flux of energy of its food—we readily agree to that. But a man composing a message on the telephone, who is apparently the source of information, poses a completely different problem. One cannot explain the information contained in the message by "information" (in the sense of "homogenous order"), nor by the organized chemical energy contained in his food. If we connect a supplementary electric motor to a telephone, we change the background noise; for example, we make it more acute; but we do not make it speak intelligently. This change of background noise, like the local heating of the metal bar in the first example, can give me certain information—the information that a new motor has been connected. In the same way, the Doppler effect informs astronomers of the radial speed of the stars or the nebulae. But in order to explain the infinitely more complex information of a message, we must have recourse to an order of the same or similar degree of complexity, to a structured, not a homogenous, order. The coupling "field of fluctuations + homogenous order" gives only a "field of fluctuations at a different level." This coupling cannot provide the equivalent of a complex modulation. The same goes for the principle of relativity. Its significance is very limited, and the invariants that it leaves intact are the essence of the universe. According to the movements of the observers, one sees, at the end of a train, a red carriage where the other sees a green rhombus, but the order and the structural arrangement of the train and of the travelers is the same for all the observers. The difference of their movements cannot transform a Mountain* into the Pacific,* or a male traveler into a female traveler.

[IV. It is true that the physicists mentioned, notably Schrodinger and Blum, claim to explain structural information itself by some "mutations" or fluctuations retained and reproduced indefinitely. The flow of force the sunlight constitutes explains, by energetic coupling, the endothermic synthesis of carbohydrates. However, the bombardment by the most energetic photons, over millions of years of evolution, would explain the progressive constitution of protoplasm capable of utilizing sunlight. This time, we cannot reproach this theory for being beside the point of the problem. However, it rests on some flagrant errors, both biological and psychological, and on a quite rudimentary philosophy. I will simply mention here the biological errors that we have discussed at length elsewhere[8]: genes, mutated or not, are not "patterns" ["*clichés*"] that transfer their order to the macroscopic organism through impression. They most likely act through the intermediary of hormones, or substances with an analogous role, since one can suspend their action with other chemical substances. They act as modulators of a given form. It is more than doubtful that genetic mutations are the unique key to the evolution of species. On the other hand, the biologist physicists are unaware of the most certain results of experimental embryology, which have revealed the extent of possible regulations in development, and the thematic character of this development.

We can further expand on the psychological errors. Psychological invention, the creation of information, certainly has nothing in common with Brownian motion, either closely or remotely. In the famous passage where Henri Poincaré describes the psychological circumstances of some of his mathematical discoveries, he uses a playful metaphor, comparing the mental agitation that precedes invention to the Democritean chaos of hooked atoms.[9] But it is clear that this semi-humorous metaphor does not coincide with the details of his own account. The important point is always the presentiment, or thematic apperception of isomorphisms. "When we arrived at Coutances, we got into a break to go for a drive, and, just as I put my foot on the step, the idea came to me, though nothing in my former thoughts seemed to have prepared me for it, that the transformations I had used to define Fuchsian functions were identical with those of non-Euclidian geometry." "One day, as I was walking on the cliff, the idea came to me, again with the same characteristics of conciseness, suddenness, and immediate certainty, that arithmetical transformations of indefinite ternary quadratic forms are identical with those of non-Euclidian geometry."[10]

It is clear that the essential thing is the putting into correspondence of some ideas according to some thematic isomorphisms, and not the agitation of mental "atoms" in the unconscious.]

Psychological invention, like the development of biological differentiation, goes from meaningful theme to meaningful theme. The man who improvises a message on the telephone first has a general idea of what he wants to communicate; this general theme evokes linguistic habits that are themselves abstract, which control the phonetic effectors and the specialized memories of the vocabulary, in the same way that embryonic locomotor rhythms precede and envelop the more specific reflexes of the limbs. Psychologists' efforts to capture invention at its source, for example to capture the birth of a hypothesis, have failed. But we can capture more easily the minor invention that is the recovery, by a conscious being, of an altered or degraded information. [A machine, we have seen, is incapable of this: there is no automaton to correct printing errors.] If a record has a material fault, this fault is aggravated on each listening. But the listener, attentive to the meaning, mentally corrects the fault. When an old manuscript has been frequently recopied, by copyists who are unintelligent or who work mechanically, the faults usually tend to worsen, just like the faults of the record, although an ingenious philologist may succeed in recovering the original text. This psychological regulation, just like biological regulation, is linked to the thematic character of the true information. The apprehension of meaning, enveloping and dominating the subordinated material, is independent of the latter's imperfection: the relation envelops the "correlates," and can reconstitute them. The informed being, as well as the informing being, participates in the domain of meaning, as the organism does in the domain of specific themes, and that is why it can correct the letter of a message. Mechanical communication has no other role, as a tool in human activity, than to aid the receiver in finding the meaning by himself. It is for information as it is for imitation. *B* cannot understand a message from *A*, just as he cannot imitate *A*, unless he is capable, or *almost* capable, of inventing or elaborating an analogous message himself. This "almost" is the secret, not only of the communication but of the indefinite progress of information in the real world. The trans-spatial world of essences and themes appears in order to complete the incomplete spatial system. What is implied by meaning, but is not actualized, actualizes itself freely.

From the incomplete information of a crossword puzzle, which is sometimes misleading, I am able to fill the empty squares and remake the model unknown to me, without making any mistakes. These three terms: $\frac{\text{Green Blood}}{\text{Grass}}$, immediately evoke the fourth term. …

We can easily imagine a machine capable of solving arithmetical or geometrical problems of proportion, because the "meaning" is materialized by the human construction of its mechanical connections. It is difficult to imagine a machine that does crosswords. Now, invention in general can be considered as a recovery of information according to a "proportion" of meaning. As in crossword puzzles, invention consists in completing a system, glimpsed in its general meaning, from fragmentary data. If the system did not exist in some way beyond space and time, invention, the increase of information, would be impossible. If, beyond some *visibilia*, there were not some *invisibilia*, some essence analogous to the solution of a crossword, invention would be not only unknowable psychologically—which it is—but inconceivable. If it is unknowable psychologically, it is because the conscious "I" is not truly the creator of the invented information, and in a sense, it receives it like a gift from another world. However, if it is conceivable and effective, if humans have created new organic compounds, even new atoms; if they have made countless new machines; if they have found routes to the new world of technology and works of culture, it is because this new world *was there* to be explored.

The role of chance is not negligible in invention, but it has been exaggerated. Even in the use of spontaneous processes of trial and error, in man and animal, chance plays a much smaller role than was first thought. Chance only counts if it is harnessed. To resume our previous example, fluctuations at the level of grains of gravel under the action of a downpour only produce a lasting effect if there are some steps on a stairway to receive the grains. What serves as "steps on a stairway" in invention can only be a "pre-structuration" glimpsed by consciousness in the world of possibilities. Without any *harnessing* of the fluctuations, these fluctuations not only do nothing but literally are nothing. To rely on pure fluctuations to create the conscious being of meaning is therefore a contradiction, since fluctuation only creates that which is relative to a preexisting conscious intention.

If we must not confuse pure fluctuations and harnessed fluctuations, we must also not take this as an opportunity to return to banal idealism. It is not human consciousness that brings artistic or technical themes into existence, any more than it truly creates the work. Human consciousness is the *medium* between the world of possibilities and the world of things. The randomness of kaleidoscopic combinations only helps my invention because I participate in their aesthetic possibilities, because I am drawn toward them. A sort of resonance informs me that I can "harness" [*clicher*] a valid fluctuation. From then on it will live, as an aesthetic theme, with that life that Henri Focillon and Jean Bayet have well described, carried from consciousness to consciousness in the history of culture.

Some young animals, playing with some wooden sticks of different length, could, by an extraordinary chance, stumble on the motif of the Greeks, like an Athenian potter; but as their consciousness is only in relation with the organic themes of their species, no resonance will occur. For them, there isn't any difference between

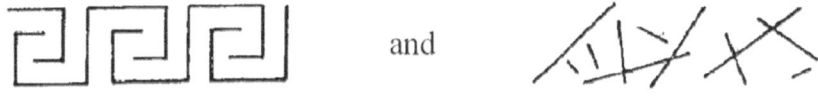

Figure 5.2.

They will only eventually be sensitive to the fortuitous constitution of a form that happens to correspond to the "gnosis" of an instinct. For a consciousness, the essential thing is the field of possibilities that it "covers." Sooner or later some corresponding works will be made: the field of possibilities glimpsed will be converted into new information. Wiener remarks that it was futile to make the American atomic bomb a secret to their rivals the Russians, because the general knowledge that the bomb is possible, since it has already been realized, was under the circumstances the decisive information.[11] This quite correct remark can be turned against the cybernetic conception of information. An inventor who explores a new domain does not possess, by definition, decisive information of this area; it will only advance if his efforts succeed. But the possibility that he glimpses in the real world plays more or less the same role as the knowledge for Russian scientists that the atomic bomb was certainly feasible, because it had been made. The glimpsed possibility plays the same role, simply with more risk of error.

The poor psychology of mechanistic theories of information and invention is exacerbated by poor philosophy. Physicists do not have the right to speak of "conformity to possibilities," and if they take this right,[12] they are no longer mechanists. The intuition of possibilities is the key to the problem of the origin of information. But this intuition is characteristic of consciousness and its relation with a "trans-spatial." Admitting this is therefore to give up on explanations based on mechanism and chance.

Contrary to widespread belief, the hypothesis of "harnessed" ["*clichées*"] fortuitous fluctuations, as the origin of structures and of information, is an inconsistent hypothesis.

Chapter 6

Negative Anti-Chance and Positive Anti-Chance

There is a close relationship between the increase of information and the presence of a domain of consciousness. Consciousness, that is, the apprehension of essences, and their conversion into well-connected [*bien liées*] actual forms, is positive anti-chance* par excellence, although all connections can play the role of anti-chance. But let us first describe what a negative, or "pure" anti-chance would be, starting from the analysis of irreversibility.

Mechanical phenomena are reversible. If the rectangle R, made up of squares a and b, represents a billiard table with perfectly elastic cushions, a ball going from *a* to *b* can return from *b* to *a*, and vice versa. The same applies if the rectangle represents a receptacle containing a very small number of molecules, all initially concentrated in a. They can go and come back. But if the molecules are very numerous, the law of large numbers makes this general return very unlikely. At usual pressures, and with a number of molecules close to Avogadro's number, the fluctuation in the number of molecules in a and the number of molecules in b is insignificant relative to the total number of molecules. To attain a fluctuation of one part in a hundred

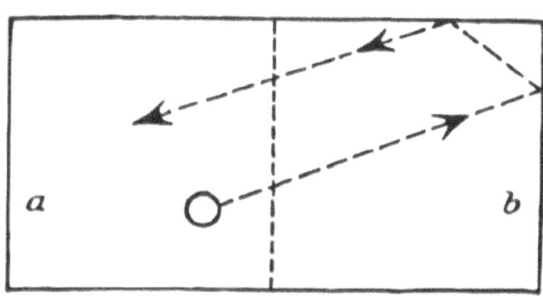

Figure 6.1.

thousand—which is therefore extremely small and difficult to observe—the probability is only in the order of 1 over a denominator with a million million zeros.

In Brownian motion, reversibility is achieved because the particles whose agitation is observed are very small and therefore receive at each instant a limited number of molecular collisions, which are distributed unevenly on their different sides. The particles push the molecules and are pushed by them indefinitely. Their kinetic energy is transformed into molecular agitation, and vice versa. The particles of Brownian motion are, in a sense, large molecules that participate directly in molecular agitation, and have the same average kinetic energy as the molecules around them. A motorless balloon in the air can be considered similar to a large particle. Theoretically, it is possible that one of its sides could receive more molecular collisions than the other for a significant period of time, causing it to move forward without a motor. It would then take its speed directly from the kinetic energy of the molecules that are hitting it. [These molecules would slow down, which would cool the air behind the balloon and leave a trail of cold air behind. There would be nothing contrary to the law of the conservation of energy, since heat would be transformed into motion.] However, according to the calculation of probabilities this phenomenon is impossible: the probable deviation required for its production has only one chance in an incredible number of contrary chances, so it is entirely excluded. In the cylinder of a locomotive, the very fast molecules that hit one side of the piston are not balanced by equally fast molecules on the other side, so there is no need to rely on an improbable fluctuation of pressures to operate the piston. The steam pressure results from a homogeneous order, brought about at great expense by the boiler through a chemical energy expenditure whose distant origin is the ordered radiation of the sun.

ANTI-CHANCE AND BACKWARD FILMS

A reversible mechanical phenomenon becomes irreversible as soon as the law of large numbers intervenes. Let us suppose two balls, *a* and *b*, with ball *a* striking the stationary ball *b*. If we ignore the effect of rotation, *a* will remain stationary after the impact and *b* will move with the same speed that *a* had before the collision. But one can choose those points of reference arbitrarily. There is nothing that prevents us from describing the same phenomenon by considering *a* as initially stationary, and *b* as initially in motion. It is also possible for me to conceive that I can make a ball retrace its path at a given moment by imparting a speed that would be the exact opposite to its current velocity. Everything will happen as if the motion had been filmed and the film were being played backward. The projection of the film in reverse will simply amount to a change of reference points: "*a* colliding with *b*" becomes "*b* colliding with *a*." The laws of collision will be respected in both the forward and backward films. The reversal of time will simply amount to a change of spatial reference points. Moreover, the reversed film will not disturb our reason: it will present us with a phenomenon that is both familiar and natural, and it will not make anything appear miraculous.

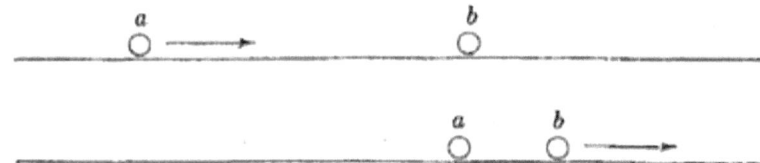

Figure 6.2.

Let us now consider a large number of stationary white balls in section *b* of a billiard table. We shoot a red ball at high speed from section *a*, while filming the operation. The white balls that are struck first will in turn strike the others, and a general agitation will be made in which the red ball will have no particular role. If the collisions were not damped, and if the cushions and balls were perfectly elastic, the agitation, spread throughout the entire billiard table, would last indefinitely. Let us now play the film backward. We will see the agitation of the balls first; it will seem completely normal to us, despite being reversed by the film. But at some point, we will see all movement concentrate on the red ball, while the white balls become stationary again as a result of the imparted collisions. In the end, the red ball will speed toward section *a*. The case of the red ball will be entirely analogous to that of the balloon [and, as with the balloon, its motion will presuppose the "cooling," that is, the immobilization of the white balls, which is equivalent to the decrease in thermal agitation of the air molecules behind the balloon]. Additionally, at the end of the film, all the white balls will be concentrated in section *b*.

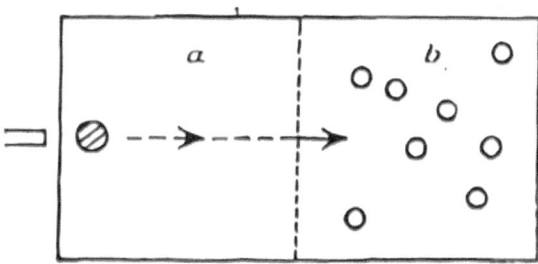

Figure 6.3.

The reversed film, although compliant with the law of conservation of energy, will not appear natural or rational this time. It will seem to produce something *ex nihilo*. The change in the direction of time—brought about by the reversal of the film—cannot be considered equivalent to a change in the spatial reference point. Each white ball, converging toward a rigorously determined point, will appear to require a well-suited system of references. The convergence of the movements will be a gratuitous phenomenon, similar to the miracle of the fortuitous appearance of the Aeneid. Contrary to what a disciple of Hume would argue, it is not only our mental habits that will be challenged. It is our reason that will be disturbed, or would be, if it were ignorant of the artifice of the film. For it would be forced, in the face of the phenomenon, to suppose the occurrence of an extremely improbable event, contrary to Bernoulli's law, which would be similar to the falling of a huge number of sixes in a dice game or of heads in a game of Heads or Tails. This supposition, as long as the extremely improbable event (that is, the indefinite fall of sixes or heads) continues, quickly amounts to a violation of the principle of noncontradiction. Indeed, the fundamental law of probability calculation, namely the law of large numbers, is based on pure abstract combinatorics; it is confirmed by the facts, but it is not derived from the facts, contrary to a fairly widespread belief. [It is enough to mathematically calculate the possible combinations to see that, for eight rounds of Heads or Tails with three successive tosses, the cases of mixed series outnumber the non-mixed ones by 6 to 2; for sixteen rounds of four successive tosses, by 14 to 2; for thirty-two rounds of five successive tosses, by 30 to 2, and so forth. Pascal's arithmetic triangle, obtained very simply by successive additions, gives the possible distribution of heads or tails. The bell curve, or binomial curve, is derived from the arithmetic triangle. It can be seen that the predominance of mixed results over homogeneous results rapidly becomes overwhelming with the number of tosses. Therefore,] after having established the mathematical proportion of the various possible combinations (by a calculation as simple as two plus two equals four), to admit the anti-chance at the origin of the continuous falling of heads is to affirm and deny at the same time the assertion "The mixed combinations are overwhelmingly predominant."

When a die falls indefinitely on the same face, we look for the reason in a dissymmetry of the die, or in the hidden functioning of an invisible mechanism. In the hypothetical backward film, we could not find the reason for the dissymmetry in the movements of the billiard balls. We would be in the presence of pure anti-chance, in the negative sense of the term. Pure anti-chance would mimic the effect of a system of connections without actually and positively bringing about this system of connections between the various movements of the balls.

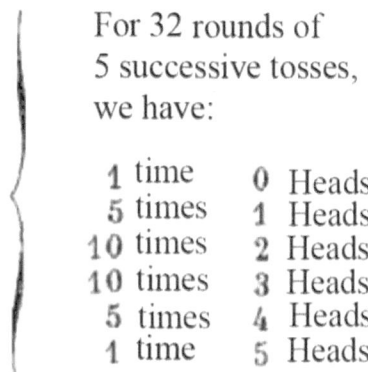

Figure 6.4.

Indeed, if one manages to maintain the original order and arrangement of a phenomenon through connections, its reversibility is no longer rationally disturbing for reason. If I drop a stone, it has no chance of going back up without a new expenditure of energy because the overall movement of its molecules has turned into disordered movements. But if the stone were attached to a rope and lowered by a pendulum motion, it would go back up naturally because, thanks to the rope, the overall order of the movement of its molecules would be preserved. It would be the same if the stone were attached by an elastic band, or if it could spin while winding around a string like a yo-yo. If the red ball in our example were attached by an elastic band to the rail on side *a*, it would obviously return to it without any miracle. Connections maintain organization. *Connections are the only rational anti-chance, the only one we can rightly speak of.* But we still need to distinguish between two cases:

a. In the first case, anti-chance is simply conservative: it prevents homogeneous order from transforming into disorder and mixing.
b. In the second case, anti-chance is truly positive; it creates order. Not only does it stop the increase of entropy, it also increases information. In both cases, there are always connections, but of different kinds. It is obviously crucial to understand what these two kinds consist of.

  a. *Conservative systems of connections.* Most machines kinematically preserve information, extend it (without increasing it), or transform it in a standardized manner according to their articulations, using effects such as rails, sliders, cams, and so on. Machines that utilize various resonances do not essentially differ from kinematic machines. Machines can also dynamically preserve homogeneous information or order, for example through elastic bands [*liens*] or fields of force. Self-distributing systems capable of self-regulation after disruptive events, which have been considered as the key to the natural emergence of order from disorder, are nonetheless pure and simple machines. From this point of view, they create neither homogenous order nor information, properly defined. Such systems dynamically *recover* the original order, or the order that is inherent to them, but they do not increase it. If we shake and mix water and oil, a homogeneous order is spontaneously reestablished, with oil on top of the water. But there is nothing more extraordinary in this than the return of the red ball if it were held by an elastic band. Gravity and difference of density here play the role of a dynamic connection. In a free-falling elevator, water and oil would not reorder themselves. In psychology or biology, attempts to explain invention through dynamic regulations are reminiscent of this horrendous mistake of believing that an increase in entropy can be, in and of itself, an increase in information. This mistake is no different from believing that a perpetual motion machine can be made using elastic bands and indefinite returns to equilibrium.

An interesting case of a machine that conserves order or information is that of the tube [*tuyau*]. It is surely no coincidence that organic machinery makes such extensive use of it. A mammal contains hundreds of kilometers of them, and Lichtenberg was able to say that all essential vital functions use tubes. It can be added that the pipe [*tuyau*] is of crucial importance in any civilization. Among other properties, the tube can combine kinematics and dynamics, articulations and elasticity, rail or slide effects, and *Gestalt* effects.

b. *Systems of connections that increase information.* These create structured order, and it is clear that they must be of a different nature and involve something other than the conservative systems of connections that operate in space-time. Let us first examine an intermediate and controversial case, that of memory and mnemonic consciousness. Someone gives me a deck of cards neatly arranged by color, and I shuffle it thoroughly. I can then put it back in the original order. Someone provides me with lines of coins, all carefully arranged heads-up. I play with these coins, flipping them heads or tails. After a while, heads and tails are equally likely. But I can then flip all the coins that fell tails and reconstruct the originally ordered lines. While it would have taken a very long time and an extremely unlikely chance to return me to the starting situation, thanks to mnemonic consciousness the restoration of the original order is almost instantaneous. It is precisely this time-saving role of mnemonic consciousness, which is also that of consciousness in general, that manifests most clearly (if not most deeply) its anti-chance character.[1] In rearranging the cards or coins, I am not strictly speaking reversing the evolution of entropy. I expend as many calories in performing these reordering operations as I do in flipping heads or tails, or in shuffling the deck of cards. I expend as much, but there is no reason to suppose that I expend more. The energetics of the phenomenon are not important here.

If one wishes to understand this phenomenon without abandoning the mechanistic principles and the connections examined in the previous sections, it will be necessary to resort to a theory of mnemonic consciousness that reduces it to a kind of cerebral elasticity. The cards that scatter, and the coins that fall at random, would be analogous to the red ball held by an elastic band, and returning to the side from which it came. The memory of the original order, materialized in my brain, would explain the return of the original order, as if guided by invisible elastic bands.

This crude and implausible conception may contain a small grain of truth, because an organism is *also* a machine, and includes nervous feedback or physiological *Gestalten* that can indeed play a certain role in particularly simple mnemonic regulations. But it does not go very far. Indeed, mnemonic consciousness imperceptibly merges into inventive consciousness. Now, in the case of invention proper, how can the supposed elastic bands cling to a cerebral physiology or to a dynamism of the classical kind? The tension felt by the inventor, his dynamic orientation, is very real, but it necessarily takes place not between elements in space—which by definition are not given, since it is a matter of invention—but between an incomplete spatial structure and an active hypergeometric "structuring" and informing from another "world." Positive and creative anti-chance does imply a system of links [*liens*], but between two worlds. The specific role of consciousness is not so much, negatively, to save time from endless random fluctuations, as to be a dynamic intermediary between the two worlds—that is, the world of meanings, essences, and themes, and the spatial world of incomplete structures. Even if I have never seen cards arranged, or lines of coins all turned heads-up, since my consciousness is an "absolute survey," and frames the actual with possibilities, it grasps the resemblances of colors and figures, and the idea "comes to it" to gather similar figures and arrange the coins—the same way a child would get the idea while playing. In figure 6.5, where the two "hearts" are mixed with different figures, I immediately grasp their resemblance, despite their segregation. Their physical distance is not physically removed by conscious "survey"; it is not overcome by a process of influence or resonance, or with an elastic band. The figure on the right is not automatically applied to the figure on the left. The two figures remain where they are, and yet they are seen as similar; they are "linked" to the same a-spatial meaningful schema. And *as a result*, they become virtually associable in a physical arrangement. The invention that consists in imaginatively arranging figures according to their resemblance is not of a very high level; it resembles a functioning. Thus, one can conceive of a "perception" machine, capable of arranging cards according to their shape or color, or even a machine capable of remotely reproducing a template shape. But the operation of this machine would have a very different mode from that of the modest psychological action it would imitate. Modern industrial workstations, equipped with photoelectric cells and thyratrons, reproduce a template shape only by exploring it step by step. Consciousness grasps a resemblance, and consequently acts with regard to this resemblance, independently of any spatial process. Even if one of the two hearts were incomplete, the "surveying" consciousness would be able to complete it by putting it in relation with the a-spatial schema that is its meaning.

CONNECTIONS AND CONSCIOUSNESS

There remains a difficulty. If connections, whatever their nature, provide the positive element of anti-chance, they must have common characteristics, and it is unclear what relationship there can be between mechanical connections and conscious connections other than a metaphorical one. As long as these relationships are not clarified, one will always be tempted to give priority to mechanical connections and reduce conscious connections to them. This temptation must be cut short. Of these two types of connections, it can be shown that it is the conscious connections that come first.[2]

[There are only two possible philosophies of connections in relation to consciousness. Either consciousness is only an accompaniment to the play of physical connections, and it is ineffective. Or, consciousness is the connection itself, and therefore, where there is binding effectiveness, there is consciousness or "absolute survey." One cannot have two different standards. There is no conceivable middle ground between epiphenomenalism and—what we

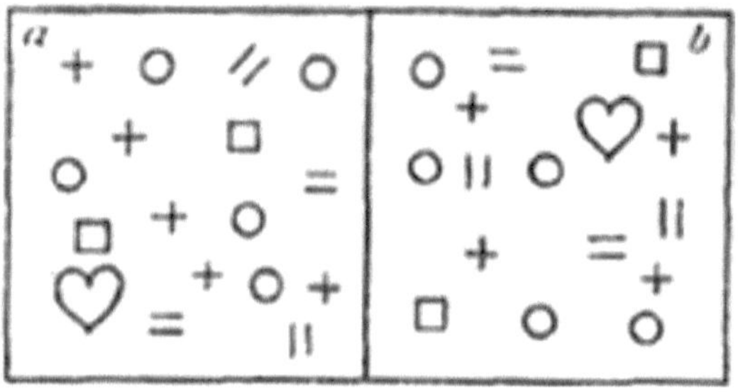

Figure 6.5.

will not call pan-psychism, for the word has taken on too-fantastic meanings, but rather the theory of consciousness as the essence of all connections and as a universal "informant." Intermediate concepts, whether in biological organicism or in the finalistic behaviorism of psychology, are only vain subtleties hiding a profound indecision of thought and the internal contradiction of a "square circle."

Clark L. Hull, for example, following Edward C. Tolman, believes that he can reconcile the recognition of the organism's de facto finalistic behavior with a behaviorist and cybernetic system. Regarding consciousness, he writes, "Is its existence denied? By no means. But to recognize the existence of a phenomenon is not the same thing as insisting upon its basic, *i.e.*, logical, priority. Instead of furnishing a means for the solution of problems, consciousness appears to be itself a problem needing solution."[3] However, he does not realize that claiming to explain consciousness is to deny it, or to make its existence appear as a vain glimmer. Consciousness cannot be explained. One can deny it practically by granting it a mere ghostly reality—or, on the contrary, by making it the effective principle of all connection and information.]

MNEMONIC RETURN

The priority of conscious connections already appears in the intermediate case of memory. Although memory, whether organic or psychological, does not increase information, it reconstructs a form according to modes that cannot be reduced to the action of mechanical or dynamic links, to the action of elastic bands or tubes, in any form. The return in space of a specific adult form, starting from a germinal cell, cannot be explained in the same way as the return of the red ball held by an elastic band. [This is because the structure of the germinal cell is simpler than that of the adult form.] No "string"—represented by a physical tracing or the action of a special chemical substance for each structural detail—can unite, one by one, the structural details of a cell and those of the adult organism composed of billions of similar cells (in terms of their chromosomes), to the germinal cell. Despite the reluctance of biologists, it is becoming increasingly clear that mnemonic return does not take place in our space, but from a world of types *toward* our space, along a hyper-geometric dimension. Embryological regulations, which are so extensive and striking since they succeed in making a normal form even after cutting, disruption, deficit, excess, or transplantation, are evidence that regulatory information is at work in development. Now, it is perfectly certain that this cannot be mechanical feedback, involving recurrent circuits and conductors located in space, since by definition, such circuits do not yet exist in an organism that is precisely in the process of making the assemblies that will later serve for its adapted behavior, and which does not yet have a nervous system. It is no less certain that it is not dynamic feedback either, using fields of force and not circuits, since experiments have condemned Vogt and Goertler's hypothesis, as well as gestaltist interpretations of organic regulation. It therefore remains the case that it is a matter of regulations analogous to those of axiological feedback, controlled by a trans-spatial ideal. The regulatory ideal in biological reproduction is mainly mnemonic: the new organism does not, strictly speaking, invent its form, which is that of the species. It is only capable of those small inventions that are regulatory harmonizations. But it is indeed a trans-spatial ideal, and the appearance of the form in space-time is an epigenesis, like invention itself.

Something similar to this return of type, from the trans-spatial world to space, also occurs in psychological memory, since redintegration from an evocative fragment does not occur in a "domino game" as was previously believed, but through the perception of meaning or a certain expressiveness.

RETURN OF TYPE AND MICROPHYSICAL BONDING

While it is impossible to understand mnemonic return of the normal type by invoking mechanical guidance, the opposite is not only possible, but is necessary, according to the most certain results of contemporary science. Every bond [*liaison*[4]] or elasticity must ultimately be interpreted as a "return of type."[5]

[Despite the complications due to various a-structured bonds in metals, elastic resistances and cohesive forces are reduced (according to W. Rossel and M. Born) to electric forces, identical to those involved in chemical reactions. The same is true, according to William Astbury, for the elasticity of organic fibers (for example, myosin or keratin), where it is the HN-CO affinity in the protein chain that sometimes closes, and sometimes, when the fiber is stretched, allows the ring to open ... CO—CHR—NH—CO—CHR—NR ...

The electrochemical bonds, whether hetero-polar or homo-polar, are reduced to a "tendency towards redintegration" or "reconstitution of a type." The Na ion, which is electropositive due to the loss of an electron, attracts the Cl ion, which is electronegative due to the gain of an electron. But why does the chlorine atom tend to capture an electron, and why does the sodium atom tend to lose a peripheral electron, if not because the normal *type* of the M-shell is to have eight electrons and not seven or one? Similarly, in the chlorine molecule, each of the two constituent atoms shares an electron in such a way as to reconstitute the structure of a noble gas (whose outer shells are "normal"). The two chlorine atoms remain linked because they tend to

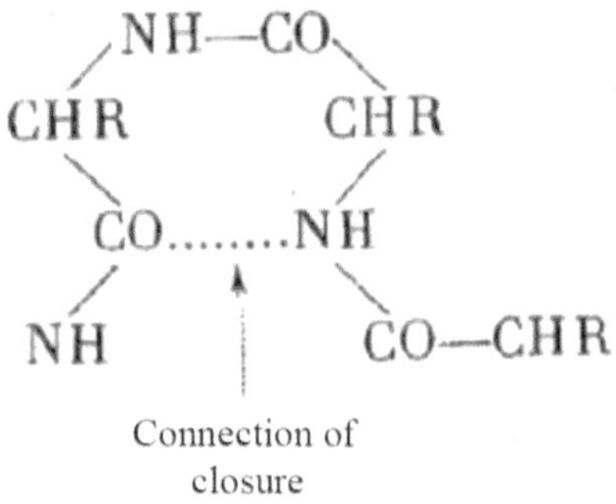

```
            NH—CO
       CHR        CHR
          CO......NH
        NH          CO—CHR
           Connection of
              closure
```

Figure 6.6.

separate either along the *a* or *a'* lines, or along the *b* or *b'* lines. Separating along the *aa'* lines, one of the two atoms would be normal with respect to the outer electron shell, but it would be in electrical imbalance. Separating along the *b* or *b'* lines, both atoms would be in electrical balance, but they would be abnormal with respect to the completion of the electron shell.

It is very curious that the formation of molecules of various types—and therefore the variety of substances in the universe—is made possible by the discordance between the requirements of two "typifications": electrical balance between nucleus and periphery on one hand, and filling of the peripheral shells on the other. Generally, when the "electrical" type is satisfied, the "shell filling" type is not, and vice versa. When there is coincidence, as with the noble gases, the complex structuring of substances is immediately halted. On the other hand, the possibilities of complexity are maximized when the antagonism or "deficiency" [boiterie] between the two types is most pronounced, that is to say when electrical balance is only achieved when filling is at half its value. This is the case with semiconductors, intermediate substances between metals and metalloids, such as carbon, silicon, and germanium, substances whose chemical activity is at its maximum. Douglas Henderson and Lesser

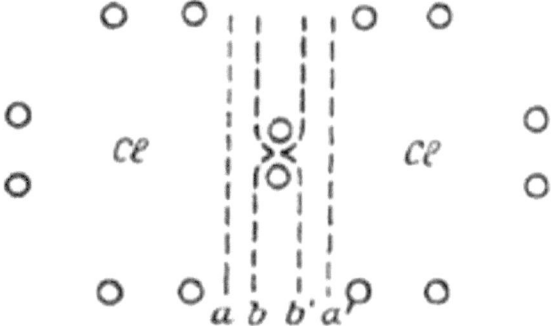

Figure 6.7.

Blum would really have some justification here for seeing a curious fitness in the architecture of things, without which the complexity of molecules, and therefore of organisms, would be impossible.

A perfectly analogous phenomenon is observed in the superchemistry of atomic nuclei. The antagonism, the "deficiency" between two types, is found in another form—this time, it exists between nuclear forces on one hand, and electrical forces on the other. For example, considering nuclear forces alone, a carbon nucleus would immediately capture an *alpha* particle and become an oxygen nucleus, since the transmutation would take place with loss of mass and release of energy. But the electrical repulsion between the alpha particle and the carbon nucleus prevents transmutation, except in rare cases where the speed of approach is sufficient. Without this electrical repulsion, carbon, as well as oxygen and all distinct simple substances, could not exist, and the entire universe would quickly be reduced (as it may have been in the beginning) to a single nucleus. For the formation of atoms as well as molecules, the shift of electrical and other forces plays the role of a sort of stairway, which breaks the continuity of the slope and allows for the existence of distinct substances.

In any case, the formation of atomic and molecular structures results from a sort of effort toward the reconstitution of the normal type. One can challenge physicists to explain or describe chemical bonds without invoking, in one form or another, the ideas of "normal," "type," "typical redintegration," and "tendency," or without using a conditional verb (which amounts to the same thing). It is doubtful that this challenge can be met, because there is no functioning here *according* to a structure, but rather a reconstitution, or tendency toward reconstitution, of a structure, as in organic or psychological memory or invention.

The interpretation of valences by wave mechanics does indeed shift the mystery of typical redintegration. For example, the filling of the atom's peripheral shells is reduced to a case of standing waves. But the mystery is displaced without being cleared up. The sharing of electrons in a molecule certainly does not resemble the fitting together of particles as it appeared in the early Bohr or Langmuir diagrams. The electrons are not localizable in their "octet" and their sharing represents the formation of a common region of probability of presence. The electrons lose their identity, and a certain energy of exchange or interaction appears, borrowed from the individual energy of the constituents. This energy of interaction can be imperfectly represented as an energy of resonance: if two pendulums react to each other, their coupling creates a "combination frequency." But this is only an image whose inadequacy is recognized: the energy of exchange is unrepresentable in our space. The "combination frequency" is an indescribable periodic change of state, where some quanta of action that are fundamental, and have no analogue at the scale of ordinary physical phenomena, always intervene. Therefore, one should not be fooled by the geometrical schemas of the atom and the molecule. There is no simple and familiar way to represent the typical structuring of the atom and the molecule. When we have a container of a determined shape and a limited number of non-deformable balls, the various ways in which the balls can be arranged in the container are also limited in number, and the resulting figures are, in the broad sense, typical. There is something coarsely analogous in the structuring of the atom. The filling of the electron shells as peripheral electrons are added superficially resembles the filling of a container with non-deformable balls. But we are sure that this image is deceptive. The particles are not balls. And the shells of the atom are not containers. The interpretation of Bohr's orbit-trajectory as a standing wave orbit (where the length of the orbit must be divided into a fixed integer number of wavelengths) should even less suggest a simple filling phenomenon. In the atom, there are no available *places*, only available *states*, calculable according to the possible combinations of the four quantum numbers. The electrical or magnetic filling of an atom is not a spatial filling.

The image of filling a container with non-deformable balls only becomes relatively true for the assemblies of ions determining the structure of crystals, in the phenomenon called coordination. To study the different types of coordination, the image of a ball surrounded by other balls, either of the same diameter as the first or increasingly larger, is usable and fruitful. One can successively arrange, according to their relative diameter, twelve, eight, six, or four peripheral balls around the central ball. But this type of structure is of a derivative and superficial order. Ionic bonds do not constitute true molecules but only crystals. On the other hand, the ionic bond, taken in itself and not in terms of the coordinations it determines, obviously cannot be explained in turn by phenomena of filling according to simple geometric laws.]

In contemporary theories of bonding, we clearly reach a limit with structural explanation. Explanation *by* structure presupposes a whole grammar: space, distance, proximity, individual particles, continuity, and kinematic movement, which is inapplicable *to* structural explanation. The atom is a structuring action, not a functioning structure. Structural explanation gradually regains a certain value only when we reach the crystalline scale, and already partially at the molecular scale. It is indeed true, for example, that in the water molecule the two hydrogen atoms form, with the oxygen atom, an angle slightly greater than 90° and that this geometric angle explains certain properties of the water molecule. But the structuring action of the atom itself obviously cannot be understood by yet another spatial structure, under pain of a vicious circle. This action requires a redintegration of type [that can only be conceived on the model of memory or invention.] An organism also partially functions *according* to its structure. But its very structuring is obviously not a functioning according to a structure. "Differentiating" cannot be reduced to a "functioning according to differences," which, by hypothesis, are not yet present in space. In this respect, the fundamental entities of microphysics resemble organisms. An atom is not like a self-distribution system or a mechanism at our scale, whether made by humans or not. Under the influence of Gestalt theory, it is often still believed that there is no middle ground between a kinematic mechanism with sliders and a *Gestalt*. It is believed that every structure is necessarily one or the other. This classification precisely misses the most central and important case, which is that of a structure formed according to a trans-spatial type and hyper-geometric bonds, whose model can only be found in the action of a conscious and meaningful theme. The structure of the atom "arrives" in space according to abstract compatibilities or incompatibilities, irreducible to step-by-step functioning.

DETOUR

In any individualized domain of microphysics, in any domain of primary connections where the individuality of the constituents is partly lost in the individuality of the system, experience reveals *behaviors* similar to those allowed in psycho-organic individualities by the existence of fields of consciousness with absolute survey. A particularly clear case is that of detour.[6] Let us consider three cases:

a. In the bed of a dried-up river, a rock has fallen, creating an obstacle. The rains refill the stream. The water takes a detour and goes around the rock. This is clearly only a pseudo-detour. The mass of water flows blindly and digs a detour channel by the effect of ordinary physical laws.
b. A seemingly very different case is that of dynamic detour. In a magnetic or gravitational field, trajectories are curved as if an attractive or repulsive center were exercising action at a distance. But the situation is the same as the water in the torrent. The bodies describing the trajectories go nowhere; they obey, step by step, the "molar" dynamics of the field. Actually, water molecules do not come into contact with silica molecules of the rock either; they are dynamically repelled.
c. A water supply pipeline in a mountainous region can use the laws of hydrostatics to overcome elevations, provided that the endpoint is at the same level as the starting point. Each siphon or U-shaped part constitutes a detour, which is also only a pseudo-detour combining kinematic and dynamic detours. Here also, the mass of water blindly obeys the laws of equilibrium.

d. An animal, finding an obstacle in its path, does not run into it; it turns aside in advance to avoid it and then resumes its course toward its goal. This is obviously the only genuine detour. But let us see what it presupposes. The behavior of the animal is directed according to a field where its current position, its direction, and the obstacle are given all at once, in an absolute survey. This involves the broad field of consciousness. To explain the properties of the field of consciousness by those of a hypothetical cortical field governed solely by the laws of molar dynamics is obviously a false solution. The animal, having accidentally entered the trap of an impasse, can turn back by momentarily turning its back on its goal, whereas the effect of molar dynamics would indefinitely keep it prisoner in the impasse. The water in the ascending branch of a siphon pipe can also go in a direction opposite to its final arrival point, but this is because the two water columns are in instantaneous equilibrium. The animal, on the other hand, balances not pure forces but some means-actions according to their direction relative to the goal. The direction, the meaning of an action, relative to the goal, can then dominate the vectorial sense of the momentary movement. The goal-set*[7] is not fully materialized in the brain. Even when the animal is in an impasse trap from which it cannot materially escape, the goal-set* nevertheless manifests itself objectively through repeated efforts, and subjectively through anxiety. The homoeostat constructed by Dr. Ashby, which gropes around in order to correct a faulty assembly, cannot however get out of a dynamic impasse; it does not avoid dead ends. If the detour, carried out in advance by the animal in its field of conscious behavior before being carried out in its geographic field, were only the projection of a cerebral homoeostatic equilibrium, the animal should be as calm when it has fallen into an impasse trap as when it has reached its goal—as calm as the homoeostat at a dead end.

e. However, in microphysics there is a phenomenon that constitutes a fourth kind of detour. The most well-known case is the tunnel effect or Gamow effect,[8] but it is very general. Despite a "potential barrier," analogous to the rim around a well, and located at a "height" such that it would indefinitely prevent the exit of a particle if it were similar to the particles of classical physics, the microphysical "particle" has a certain number of chances to exit, and a certain proportion of exits actually occur. It cannot be a question here of invoking a kind of siphon, or slide, or elastic link acting according to any guidance to explain the crossing of the barrier. The particle does not follow a representable trajectory step by step like water molecules in a pipe. The forcing of the barrier is excluded by the "molar" dynamic situation. But wave models make it

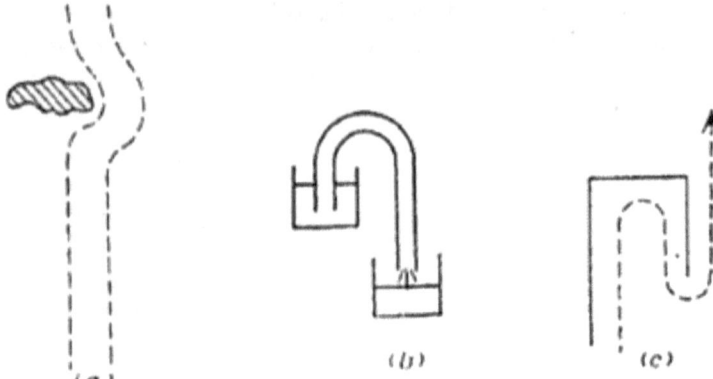

Figure 6.8.

possible to understand the tunnel effect, since waves of probability—of the particle's presence—can bypass or pass through a potential barrier.

Although the microphysical detour does not resemble the conscious detour, at least in the current state of scientific conceptions, it is nonetheless much closer to it than to the others. It is characteristic that physicists cannot avoid using terms borrowed from psychology to characterize the microphysical detour. The particle, through the probability waves that constitute it, "explores space" instead of simply describing a trajectory; it is "potentially present" in a whole domain of probability.[9] Whereas the classical particle *goes* nowhere and the end of its movement is simply the result of preceding elementary movements, the microphysical particle "takes into account" the "intrinsic energetic legitimacy" of the final state and integrates the "means" to achieve it into a unitary action where time, as well as space, seems to be surveyed, as in a consciously operated and calculated detour. The microphysical action, like the psychological action, seems to take place against a background of utopia and uchronia, a background of "Otherwise . . . " Its actuality is surrounded by "glimpsed" possibilities.

In any case, the microphysical detour and the conscious detour are clearly related. They both involve true and individualized actions. Other types of detours are degraded cases, which only appear through the multiplicity of individual actions, whether they are conscious individuals or microphysical ones. A crowd channeled in the corridors of the metro or in the streets of a city hardly differs, as far as its capacity for detour is concerned, from a crowd of water molecules channeled in a pipe. It also stupidly hits an obstacle that accidentally blocks its path. Each of the individuals that compose it calculates their walk, through conscious survey, relative to the immediately neighboring individuals, but the absence of a general survey degrades the behavior of the crowd as such.

In any case, the channeling connection [*liaison*], and the structuring it applies to amorphous crowds, cannot without contradiction serve to explain the elementary connections and structuring that emerge *de novo*. Standing waves, which *constitute* the electronic layers of the atom, cannot again be like waves *in* a sound pipe. Electrons, or associated waves, do not circulate between solid walls. A pipe, or a system of conductive slides, preserves information; it delays its degradation, or materializes information previously created by a field of consciousness, for the use of a blind crowd. The siphon and U-shaped piping was built by an engineer. Similarly, the road or railway that circumvents obstacles winds its way according to the slope, or passes through bridges and tunnels. The few natural pipelines that can be cited (artesian wells, intermittent springs, natural bridges, etc.) *are of little importance and are truly lusus naturae.*[10] It is obviously necessary for a certain primary "linearization"[11] of causality, or rather of individual activity, to occur directly by surveying and calculating possibilities, so that secondary linearizations, by tubes or slides, are conceivable without contradiction or infinite regress. If the entire domain of microphysics is precisely the area where "it is necessary to stop," and where "one can stop," it is because microphysical individualities are sources of information and structuring as primary as a field of consciousness. Structuring bonds [*liaisons*] cannot again be explained by pipelines or walls. Otherwise, another wall or rail would be needed to "linearize" the wall or rail, and so on indefinitely.

What prevents us from grasping this obvious fact is that at our scale, the movements of bodies or particles appear to us to be naturally channeled by the material consistency of the moving bodies. But movement, according to wave mechanics, is always propagation, and not material transport. The movement of a supposedly substantial moving body is only the secondary appearance of an electrical, luminous propagation, and more deeply, of a system of probability waves. A certain wave behavior must be defined before we know what is propagating and what is moving. The manner of propagating is more fundamental than "the body moving in space."[12] Therefore, this propagation of information, which is essentially movement, must channel itself, must invent its own connections in its own field. Without this self-channeling by invention, that is, by participation in a chosen type or possibility among possibilities and nonactualities, nothing could exist in our space; everything would be lost in absolute homogeneity, or zero information. In the microphysical domains, binding information must *appear*; it cannot simply be *conducted*. It appears by epigenesis, as organic or psychological differentiations appear.

If Descartes is the ancestor of cyberneticians through his physiology, through his embryology he is the father of all those who want to do synthetic morphology, that is, who want to explain the appearance of forms solely through the play of the laws of ordinary physics. It is characteristic that all Cartesian embryology is based on the use of connections of type *a)*, either dynamic or kinematic. According to Descartes, the first sketches [*ébauches*] of the fundamental organs, and especially of the heart, the fundamental organ par excellence, are due to a sort of dynamic equilibrium—*Gestalt avant la lettre*—to an "agitation" produced by the meeting of the two germs. Then, circulation shapes the circulatory system, like wild water carving its bed. And finally, the other organs are the by-product of the channels or tubes thus formed: the kidney, bladder, and ureters are the *excrementa* of the vena cava; the sense organs are the cerebral *excrementa*; they are brought about by the channels like sediments by a river.[13] As these type *a)* connections can at most only preserve information and not increase it, the Cartesian attempt, which was quite crude, was doomed to failure. But it can be said that the most modern attempts, insofar as they also rely on similar types of connections, are no more valid.

In fact, organic differentiations appear from simpler forms and cannot be attributed to dynamic regulations. The primordia that will become the circulatory system, for example, appear here and there before the flow of circulation, and then merge to form the vascular network. As recent experiments by Jolly have shown, it is even the case that the destruction of the presumptive cardiac primordium does not prevent the development of the corresponding aorta. The organism actively forms its own tubes, channels, and conductors—by definition, without preexisting tubes, channels, or conductors. Similarly, in the psychological realm, if I move in terrain where there are no roads or established paths, improvised routes and itineraries appear in my field of behavior, no doubt inspired by the never-quite-homogeneous form of the terrain, but not rigorously determined by it. The engineer who draws roads seeks the greatest economy and best efficiency, but in his creative effort, he does not passively obey the principle of least action, like a drop of water on an inclined plane. He is inspired by an ideal type that, with the risk of error, he works to realize. By seeking appropriate ways and designing various sections of the road, he takes into account the point of arrival as well as the point of departure. The same applies to the tubes and auxiliary connections in the living organism. They are never pure effects, cumulative functions or step-by-step equilibrium, or beds dug by the effect of a blind dynamism. The vessels essential for pulmonary respiration are formed in the embryo, oxygenated by maternal blood, before they become fully useful. Conversely, the vessels that are no longer useful at birth do not disappear, like the old bed of a dried-up river.

At birth, the arterial duct that, in the embryo, directed the pulmonary circulation into the systemic circulation, undergoes a physiological closure by contraction of its membrane, then it anatomically obliterates itself. The

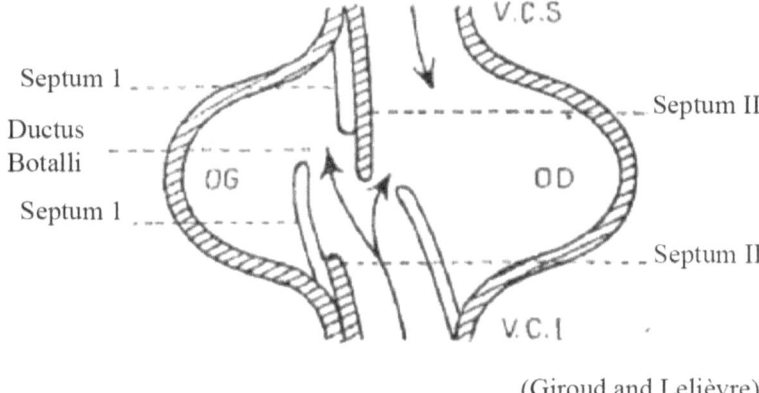

(Giroud and Lelièvre)

Figure 6.9.

ductus Botalli,[14] which communicated between the right and left heart, undergoes a physiological occlusion; then it closes, three or four months later, by anatomical fusion. The presence or absence of vessels is visibly subordinated to something other than pure "a tergo" causes. It's true that the physiological closure of the ductus Botalli seems to occur in a purely mechanical—or at least "causalist"—manner, by the effect of the increase in blood pressure in the left atrium.

The *septa* that constitute its membranes, especially the *septum primum*, function as valves and come together. But then it is necessary to explain why the *septa* were precisely formed in advance, in such a way that at the right moment, a purely mechanical effect could make them function according to the needs of the organism.

As we have seen, organic memory cannot be the result of the functioning of the organism. Once the channels and tubes, the auxiliary instruments of connection and circulation, are constituted, those that use them—humans, animals, cells, fluids, waves, or particles—can be reduced to a blind crowd obeying global impulses or forces. But to explain their very constitution by what is precisely permitted only by this primary constitution is to get things backward.

In structuring action, the action of positive anti-chance, we again find the same schema of framing as in action in general or in communication. Direct improvisation, in a field necessarily analogous to a field of consciousness, of informative connections, is fundamental, primary, and irreducible. All new information is an invention analogous to conscious invention; it is the effect of a unifying theme. Then, positive and creative anti-chance frames auxiliary devices that it has constituted and which themselves represent a conservative or channeling anti-chance: material links, conductors, guides, and so

112   *Chapter 6*

on, which function in chains of assistance and that, sometimes, completely replace the original conscious connections. All organic machines could serve as examples.[15] Conservative anti-chance then channels and envelops the amorphous crowds, composed of individuals that only react to each other step by step and gradually. If the conserving auxiliary devices degrade, pure chance prevails over anti-chance, and information decreases irreversibly if positive anti-chance does not come to reconstitute or replace them.

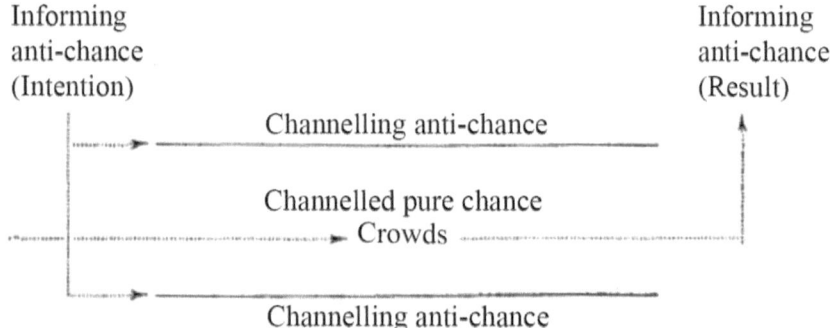

Figure 6.10.

Chapter 7

Past-Future and Cybernetics

If we could isolate the various types of connections, time would be completely different depending on the type considered in its pure form. Mechanical connections, such as sliding connections, which allow indefinite and reversible functioning, would belong to an indefinitely reversible time. Macroscopic dynamic connections belong to a time that is reversible and endlessly oscillating. Edge-to-edge connections in a crowd—that is, the absence of general connections in the crowd considered as a whole—lead to maximum disorder and belong to an irreversible time whose "limit" ["*plafond*"] is maximum homogeneity in an indefinite absence of time. Informative connections of conscious action move in an irreversible time and also lead—in the case of an individual and isolated conscious action—to a "limit" and an indefinite absence of time once the action has reached its optimal endpoint. Two points are noteworthy. The first is that no type of connection in pure form provides the time of ordinary experience, which is irreversible, continuous, and without a "limit." This ordinary time is necessarily a "compound." The second point is the curious resemblance between the time of crowds and the time of conscious action. Both are irreversible and have a "limit," extreme in one case, optimal in the other. Both have, according to Eddington's expression, an "arrow." Relative to information, these arrows always have a determined orientation, since the information is minimum at the limit of crowd time and maximum at the limit of the time of conscious action. It is their relationship that we must study now because the similarity of these two times has deceived cyberneticians.

According to Wiener's thesis, information machines, unlike clockwork machines, would exist in an authentic time, just like living beings, with a past and future that is irreversible.[1] They would exist in the time of J. W. Gibbs's statistical mechanics rather than in the time of classical mechanics. They would even exist, according to Wiener, in Bergsonian time.

It is already strange that this thesis is diametrically opposed to Schrödinger's thesis.[2] According to the latter, the conservation of information by organisms requires comparison to a molecule at absolute zero, where entropy is zero, or to a clockwork* that would function without any friction, and thus preserve its order completely.

Finally, Wiener is not perfectly clear about the relationship he establishes between the time of heat engines and the time of information automata. He brings them together, to oppose them to the reversible time of Newtonian mechanics, while also opposing them to each other. He naturally brings them together because, as we have seen, entropy and information are antithetical and symmetrical notions. He opposes them because a heat engine works by degrading energy, whereas a feedback automaton uses information in its very functioning, from "input" to "output," without "consuming" it.

THE FANTASY OF INVERTED TIMES

Wiener first notes, very ingeniously, that we cannot observe or communicate with a system other than our own unless the direction of time is the same in the observed system and the observing system. The very fact that we see a star means that its thermodynamics is similar to ours. Indeed, we perceive the light that comes toward us and reaches our eye or photographic plate; we can see stars that emit light, but we could not see stars that might "radiate backward," that is, absorb light. We know our past, but not our future. Retinas, or photographic plates, can only perform their role if they receive information emanating from the star, at a present moment whose past is not contaminated by a disorder of the past-future.[3] If we had anticipatory images, or *pre-images*, as we have residual images, or after-images,* vision, like photography, would be very difficult. If we wanted to communicate with a being whose time went backward compared to ours, it would be impossible for us to perceive their informative messages as messages. Suppose this being B wants to send us, A, a square as a message or signal, by tracing it in the sand. As our time would be hypothetically the opposite of his, we would first see at moment a for us, and b' for him, weak indistinct traces, which would then become a square, at moment a' for us, and b for him. This square would suddenly disappear for us at our moment $a,'$ corresponding to instant b where B traced it. But since we would have seen the square form itself gradually and spontaneously according to our past-future, we would attribute its formation to natural causes, and therefore it would not resemble a signal at all to us. Its sudden disappearance would also appear to us as a natural catastrophe.[4]

This fantasy of Wiener's does not exactly prove what he wants it to prove. Suppose, in fact, that instead of a simple square—whose progressive formation could, at a stretch, be seen as a crystallization or a *lusus naturae*—B sends a very complex message, comprising numerous printing characters.

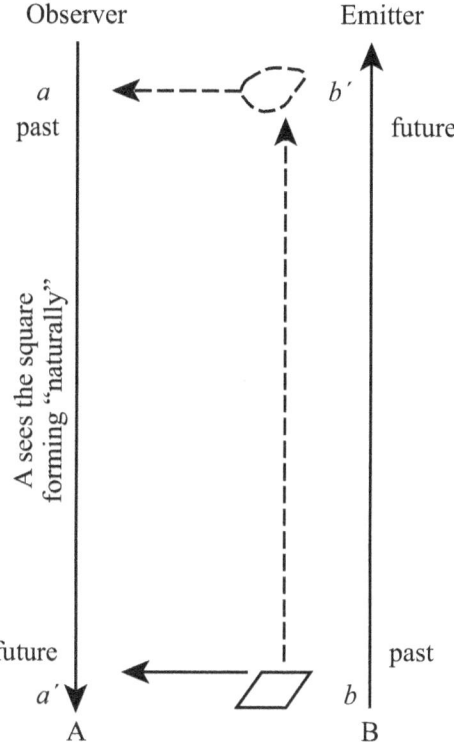

Figure 7.1.

Even in these circumstances, it would take a certain amount of bad faith not to understand the message as a message.

The fantasy of inverted times is less fantastic than Wiener imagines. Without inverting thermodynamic time, consciousness at least surveys it, and for large stretches of time is not confined to the present moment. Leaving aside the hypothesis of inversion, let us consider B as an ordinary conscious being whom we observe. It will happen very often that, from a to a', we will progressively surmise the intentions of B, beginning with a vague presentiment and ending with a precise perception. If B is hostile to us, we may sense his hostility through sign-effects, $s, s,' s',' $ which are very slight, but nevertheless decipherable to a psychoanalyst. These sign-effects, emanating from the intentions or complexes of B, in some sense go back up the stream of time for us, and even potentially for him if he is unconscious of his latent hostility. They go back up for him, or at least they emanate from a source S, timeless or super-timeless relative to the stretch of time under consideration. In invention also, it often happens that the inventor divines through sign-effects a still-hidden idea of which he has, in some sense, "pre-images" and not afterimages.* The first imperfect sketches [*ébauches*] arrive before the clear realization, exactly as, in Wiener's fantasy, the half-erased square is perceived by *A* before the clearly traced square.

A metaphysician would even be entitled to use the schema of the two inverted times—or of two times of which one is more "surveyed by consciousness" than the other—to interpret the evolution of living species according to a cosmic or theological purpose. This evolution seems to have natural causes: mutations, selection, adaptation, and so on. Consequently, those with a positivist spirit refuse to see a meaning, an idea, behind organic forms and their evolution. But perhaps it is the same illusion, due to the same inversion, or the same lag between our time and that of the Demiurge or the Elan Vital, which makes *A* interpret the square traced by *B* as a natural crystallization, or a *lusus naturae*, even though it is "meaningful."

In any case, it is undeniable that a simple change in temporal rhythm often makes it difficult to understand or even recognize a meaningful message. Conversely, a lost meaning can be recovered by changing an inappropriate rhythm. In sped-up cinema, the groping movements of tendrils, or the blooming of flowers, appear surprisingly intentional. And there is no proof that it is not the rhythm of cinema that is the "right" rhythm. If we could see a cinematic montage, made from precise documents, representing in a few minutes the evolution of man from his simian ancestors—the progressive

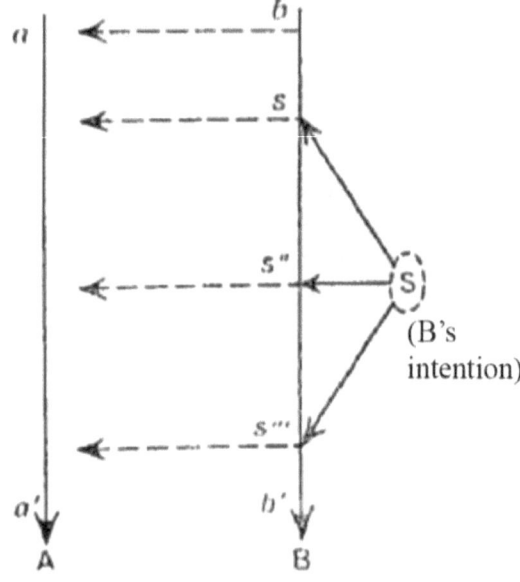

Figure 7.2.

cephalization, the upright posture from quadrupedal walking, the changing appearance of the face and gaze—it is highly likely that "meaningful" interpretations of evolution would gain many followers. Even positivists, perhaps, would attach less importance to the fact that accumulated mutations appear to explain, step by step, the appearance of man. These mutations, inserted into the overall appearance, would no longer seem to be anything more than subordinate means.

In the nineteenth century, people often believed they could erase the differences in nature between things, or sufficiently account for their specificity, by "drowning" them in a very slow evolution. It took the entire arsenal of phenomenology to escape this prejudice. Meanings and ends often appear only at the cost of a systematic disregard for details and subordinate means. Just as ruins, almost erased by sand, are unrecognized by pedestrian travelers and revealed only by aerial photography; just as there is no beauty in a painting for one who looks at it through a microscope, or there is no human beauty for Gulliver in Brobdingnag—likewise, there is no meaningful information for a Gulliver of time who adopts an inappropriate rhythm.

MACHINES AND THE PRESENT

Let us return to Wiener's thesis on the time of information machines. Nineteenth-century physiology drew inspiration from heat engines and considered organisms as natural automata, as machines that burn sugars and fats. It was the energetic balance, or the balance of the metabolism, that mainly attracted its attention. But today, physiology is mainly interested in operations carried out at a low energy level, as in electronic tubes, where what matters is not the energy balance, but the fidelity with which messages are reproduced or used. It no longer draws inspiration solely from thermodynamics but from the physical theory of communication.[5] It considers organisms as automata coupled with the outside world, not only through their energy flow, their metabolism, but through the flow of impressions, messages arriving and departing. The receiving organs of automata are equivalent to sensory organs, and their effector organs are equivalent to the muscular system. Between these two kinds of mechanisms, between the input* and output*[6] of machines, others have the function, like the central nervous system, of storing information, rules of action, and of controlling the effectors according to perception of the results already obtained. Wiener concludes that time in automata, whether natural or manufactured, has an even clearer meaning than in heat engines. It is impossible to interchange incoming messages and outgoing messages, and it is also impossible to imagine the functioning of an automaton that would have pre-images or whose memory would precede incoming messages, just as it is impossible to imagine a living being that knows its future and not its past. It is therefore clear that the input-output* relationship is in a one-way time and implies a defined past-future [*passé-avenir*] order. "Thus the modern automaton exists in the same sort of Bergsonian time as the living organism; and hence there is no reason in Bergson's considerations why the essential mode of functioning of the living organism should not be the same as that of the automaton of this type."[7]

We have acknowledged that the line of the present does indeed pass between the "input" and the "output" of information machines and, more generally, is fixed by the ongoing functioning of any machine auxiliary to an action. Only a functioning machine can fix the present with precision. It seems strange, at first glance, to draw such a metaphysical dividing line between the "input" and "output" of machines. But to convince oneself it is enough to look around without prejudice. At this precise moment, the tip of my pen, controlled by my hand, is depositing ink on my page, writing a sentence according to a psychic set-up that itself responds to the meaning I want to express and that aims for a truth. Truth, meaning, and even the sentence that wants to express it, are temporally "spread out"—truth even completely transcends time—but the tip of my pen operates in a very precise present. Cars pass in the street; every fraction of a second, a spark automatically ignites, causing the gasoline to explode and pushing the piston in order to carry the driver of the car to a specific place, responding to a psychological intention, enveloped by a larger "ideal." Birds sing in the trees of the square; their larynx operates in the rigorous present of sound vibrations according to an instinctive impulse much more liberated from time and the present, and which itself responds to some vital intention as old as the world. Even pedestrians do not move without an interposed machine. Their intention to go somewhere, to accomplish something, envelops the action of a subconscious "task" that itself envelops the physiological functions and the chronologically rigorous mechanics of a series of controlled falls. An ideal line of the present only intersects activity punctually through machines. Without machines, the actual would be a *specious present* of variable magnitude depending on the degree of embodiment of the activity.

PAST-FUTURE AND ENTROPY

It remains to be seen whether machines alone, whether they are clockwork,* heat engines, or information machines, could give, not only the present, but the arrow of time, from past to future, independently of their envelopment by consciousness. We can pose this problem in a more precise way.

Is there a purely physical criterion, really for the use of physicists, for the past and the future? To this question, physicists generally respond with a comfortable sense of security: "Yes, this objective criterion exists. To know if the instant t' is after the instant t, for example, it is sufficient to take the temperature of a hot body placed next to a colder body: if this temperature is lower at instant t' than at instant t, then t' is after t. The increase in entropy gives the direction of time." But this answer, as Eddington incidentally pointed out,[8] and Satosi Watanabe explicitly stated,[9] leads to a vicious circle. To know whether entropy is increasing or decreasing, one must already have, directly and psychologically, the intuition of the direction of time. Let us imagine a physicist who would not have this intuition, and who would only be able to measure temperatures and the value of entropy. He would therefore find different numbers at t and t'. Why, among these three hypotheses, namely that (1) "there are different entropies"; (2) "entropy is decreasing"; and (3) "entropy is increasing," would he automatically choose the third?

For his various observations to have an order number, he must already have the sense of time. He could not know if observation b, of entropy b, is after observation a, of entropy a, and must therefore be considered as the second observation, if he did not have the sense of time. Therefore, $b > a$ does not determine the direction of time. If the physicist, guided by an abnormal psychological impression—we mean, one different from ours—believed that b is before a, he would believe by the same token that entropy had decreased. He would change the statement of the law, and there would be no reason to contradict his intuition.

A first reason to think that the evolution of entropy, as we perceive it, *requires* an intuitive direction of time, and does not objectively *fix* this direction, is that this intuition persists even when we observe reversible mechanical phenomena. Thanks to this intuition alone, I can speak of a factual direction for a reversible phenomenon in principle. Newtonian or Einsteinian laws, combined with cosmological data, generally determine the direction of the Earth's rotation and therefore the apparent direction of rotation of celestial bodies. But I still need to believe as an observer that "eight o'clock" is after "seven o'clock." Because otherwise, without being surprised, I would find that the astronomical laws make the sun rise in the west, not the east. In fact, through a combination of astronomical laws and my psychological intuition of time, the sun rises in the east and not the west. To say that the phenomenon is reversible is to say that my reason *would not be* shocked if the sun rose in the west. But I put the sentence in the conditional, which is proof that, despite the permission of my reason, my sensitivity—or whatever it may be—directly feels what is past and what is future.

To know in which direction the hands of a clock go, and if eight o'clock is after or before seven o'clock, I need this intuition. It doesn't matter whether the clock is an ideal clockwork* without friction or an imperfect mechanism that internal friction makes subject to Ludwig Boltzmann or Gibbs's statistical mechanics. Because, once again, nothing would assure me that "increased entropy" means "later moment" if I did not already know it. As Watanabe emphasized, the second law of thermodynamics is indissolubly both psychological and physical, otherwise it would be a simple tautology: "Entropy increases with time." If, after that, we define the direction of time by the increase of entropy, then the principle becomes: "Entropy increases with the increase of entropy." For the principle to mean something, it must become "Given our intuition of past-future, entropy increases according to the psychological sense of past-future."

One should not imagine that the hybrid, both psychological and physical, character of the second principle of thermodynamics is a unique and exceptional case. All truly fundamental notions used by physicists are necessarily linked to conscious intuition, by virtue of the same principle that we have called the "principle of framing" of machines by consciousness. Scientific observation is most often done through an interposed machine. But one must arrive at a knowledge of what it indicates, that is, conscious sensation and perception.

The fundamental notion of simultaneity is just as much a psycho-physical "mixture" as the notion of past-future. Although this aspect of the theory of special relativity is generally overlooked, it is nevertheless indisputable. Einstein's reasoning amounts to recognizing that simultaneity is an empty notion if it is only conceived or imagined in the abstract, and if it is not realizable in a concrete observation. To say that two lightning strikes are simultaneous for a reference system is to say that if I am tied to this system, and if I observe the two lightning strikes through two inclined mirrors, I will see them together in my field of consciousness. Of course, I can be advantageously replaced by a precise machine capable of measuring simultaneity to the millionth of a second. But I will always have to read the result to give it meaning, and in any case, simultaneity will only be "realized" by the absolute survey of a field of consciousness where multiple details are both distinct and yet present, without being really "at a distance" from each other in the way bodies are in space. It is only "in" my field of consciousness that I can look at two clocks at the same time and note their synchronism, without their distance in the unique sensation having to be overcome, like physical distance, by physical means of propagation and information. If it were not so, the difficulty of defining simultaneity would be infinitely postponed and unresolved. As Eddington demonstrates, Minkowski's four-dimensional universe is entirely built on the psychological basis of the here-now, on the "seen-here-now." The whole theory of relativity consisted in abandoning the belief in a universal and abstract "now" and believing only in the concrete and psycho-physical "seen-here-now," limiting the realm of absolute past and future and separating the past and the future by an "absolute elsewhere," where events whose simultaneity, being unobservable, have neither meaning nor reality.

The error made before Einstein was therefore made due to a confusion that was as much psychological as physical. If one believed in absolute simultaneity independently of any possibility of interaction and signaling, it was because one confused the abstract concept of simultaneity with observation. Before Einstein absolute simultaneity was also reduced to psychology, but to the mythical psychology of pure thought, not to the concrete psychology of sensation—that is, to the psycho-physical mixture that is sensation or observation. As has been emphasized, the error undoubtedly consisted in radically separating time and space. But this error itself was the inevitable effect of a prior and more fundamental error, which consisted in radically separating the psychological and the physical. The notion of absolute simultaneity was based on a psychological illusion, of the same order as that which makes primitives or neurotics believe in the ubiquity and omnipotence of thought. Consciousness does indeed have a sense of ubiquity and even omnipotence, but in the limited domain of absolute survey that is its being, where there is no need for propagation from one point to another, where there can be simultaneity, action, and interaction, in a concrete, non-punctual, and yet unitary "here-now."

One should not interpret the psycho-physical nature of the second principle of thermodynamics, as Watanabe seems to do, in the sense of a purely scholastic idealism, which would consist in saying, "It is observation that creates the sense of time of the physical universe." This interpretation would be as false in this domain as in others; as false as, for example, the well-known formula "It is the scale of observation that creates the phenomenon," or any other modern adaptation of the "*Esse est percipi*." The fact that real interaction is psycho-physical and operates in a "domanial" here-now, in a monad of action, is not a confirmation of the idealistic thesis that Spirit and Knowledge create the world. Leibniz is easily adaptable to modern science, but Berkeley is not. Let us suppose that A, a conscious being, observes not a physical phenomenon resulting from Gibbs's mechanics, but another conscious organism, B. It will not then be the intuitive sense of time of A that creates the sense of time of B, since B also possesses an intuitive sense of time, and would, according to the hypothesis, have just as much right to create the sense of time of A. If A observes B still in the embryonic state, it cannot be said that he creates the sense of organic development. The paradox would be too great to dissociate organic development or embryology from behavior and psychological activity. For reasons of continuity, it must therefore be admitted that all organic, animal, vegetable, and microbial development exists in a time whose "arrow" does not depend on observation.

If A now observes B as a corpse and in a "disorderly" state according to the physical laws of increasing entropy, for reasons of continuity again it must be admitted that his observation no more creates the sense of disorganization than that of vital organization.

PAST-FUTURE AND ACTIVITY

The truth is that A's sense of time is the very sense of his informing activity and creative work. It is the psycho-organic work that gives sense to time, or at least that constitutes the raw material of the past and the future. If there were no living beings—no organisms, in the broad sense—in the universe, there would be no sense of time. This does not at all mean that it is the observation of phenomena by a living and conscious being that fixes the sense of time. It is the very fact of informative work and action that produces, or contributes to producing, the past and the future. And since it is impossible according to modern physics and biology to make an absolute break between the most basic organic individualities and molecules, one can say that all true individuals in the universe—those that still traditionally pass for physical beings as well as those that are considered organic—are activities, and therefore contribute directly to the production of the past and the future. The paradox is that when one arrives at these atoms of activity that are quantum actions and interactions, the past-future distinction probably no longer makes sense.[10]

That is why it is futile to pretend to follow in our thought, endowed with memory and psychological past-future, the supposedly progressive course of an electron on what is metaphorically called its trajectory, or to pretend to follow its very rotation. This operation is as futile as thinking of absolute simultaneity in the absence of interaction. The electron does not travel its trajectory in a "step-by-step" spatiality and an "instant by instant" or "past-future" temporality, modeled on the past-future of our actions on a larger scale. The temporal step by step of the past-future, like the spatial step by step, cannot apply to primary individuals, which do not "dominate" any "sub-individualities."

But if we consider complex individuals and, in a broad sense, "organic" ones, the sense of time results from the relationship between their intentional activity, in "absolute survey," and the subordinate mechanisms that achieve this timeless intention step by step. "Arrowed" time, as it reigns in organic evolution and psychological life, is a mixture of two elements that are in themselves non-temporal: pure thought, without physiological or mechanical machinery, which is inaccessible to us but which we approach, and which is outside duration; and pure functioning, through *a tergo* pushes between multiple elements, which exists in duration only through a surveying consciousness. It is because our intentional consciousness uses psychic sets* and machinery that it is temporalized, and even, one could add, that it is truly psychological consciousness, insofar as it is a thought that perceives itself instead of getting lost in the super-conscious transparency of essence. Consciousness exists in "arrowed" time because it is vigilance and effort. This is work and not pure act, insofar as work is always aimed at a timeless intention, that is both slowed down and pushed by the play of multiple subordinate means. Spinoza felt eternal when he thought, but certainly not when he wrote his thoughts with a bad goose quill. The cube of sugar melting in water, which Bergson takes as an example, surely does not itself experience this fusion as a duration. Perhaps each of its molecular links at the moment it unravels, but not the cube as such, which has no individuality. My desire for sweet water is also relatively timeless, although in a completely different sense. It is the fusion of sugar-enveloped-by-my-desire-for-sweet-water; it is the expectation of its automatic realization, that is truly temporalized. Time is the product of two factors: a "relative eternal" and an "enveloped machine."

Psychic sets* are, in a sense, to be classed in the category of enveloped machines. It is proper to psychological action to transform into a mnemonic state. Psychological action is an original information by the meeting of the individual with the world of essences and values; the psychomnemonic state represents a sort of capitalization of this informative work. The psychomnemonic state is an action become being, half-substantiated, and facilitating further work, as industrial and social capital facilitates social work. If we had a "memory of the future," that is to say, a direct apprehension of form by pure contemplation, there would be no need for active conversion and work, and there would be no sense of time.

Regarding the meaning of the evolution of entropy, it is not a separate problem from that of the direction of action and informing work, since entropy is a "disinformation," a negation of individualized systematic activity. There is an increase in entropy, a degradation toward homogeneous mixture, as soon as an individualized system dissolves and becomes a crowd, as soon as uncoordinated elements are no longer in systematic interaction and obey the law of large numbers. The direction of time according to the second law of thermodynamics is therefore fixed *a contrario*—objectively, and not by the magical virtue of observation—by the direction of time in individual work. Ceasing to work as an individual automatically degrades information if conservative connections do not maintain the accumulated information capital. A dead organism is nothing more than a crowd of molecules, and automatically exists in the time of thermodynamics. This time of disorganization naturally continues the time of organization, like the spreading of a colored spot on highly absorbent paper naturally continues to spread after a drop of ink falls from my pen. The time of thermodynamics, whose course is observable by the increase of entropy, is therefore not a time separate from the time of psycho-biological action, in the sense that these two times would have a separate origin, and such that we would have to look for how they can be connected despite this difference in origin. The dissipation of forms as soon as informative action ceases is correlative to active information itself. It is because I am able to see color appear that I can observe its disappearance, and it would be absurd to interpret the phenomenon by saying that decolorization requires a different time from that of coloring. It is because I am an active constituent of forms that I get the "meaning" of information, and therefore also the meaning of "disinformation."

Only psycho-biological activity, which immediately senses its own meaning,* can have the *intuition* of the direction (*richtung*) both of the information it creates and of disinformation. Only activity can experience both its successes and failures.

Cybernetics is deluding itself by believing that its automata exist in Bergsonian time[, just as it deludes itself by believing that because it encodes the information contained on a printed page it thereby enters the "meaning" of that page into the domain of positive science]. We perceive disorganization only because we are organizing activities. Consciousness is always an organizing activity, a work of organization and information, an anti-chance. Some psychological *phenomena* resemble increases in entropy (for example, ordinary forgetting, as opposed to Freudian forgetting by positive acts of inhibition and repression, although the very existence of ordinary forgetting is contested), but these phenomena are precisely not *acts* of consciousness. The proper act of consciousness is always the creation or re-creation of information; therefore, the act of consciousness gives us, by definition, a past-future. All conscious work is "arrowed," and this arrow gives, by contrast, the arrow of disorganization in the absence of work.

Films shown backward give a spectacle contrary to reason understood both as a principle of mathematical calculation and as a habit and impression of rationality. They are contrary to mathematical reason because they present phenomena that do not conform to the law of large numbers. They are immediately contrary to the impression of rationality because they present gratuitous results obtained without informative activity or, conversely, because they present activities that lead to disinformation. In a backward film, the pieces of a broken bowl re-form into a whole bowl without effort or work, whereas we know from direct experience that repairing dishes is difficult. A painter in a backward film appears to be erasing colors from his canvas, and he ends up looking confused like an amnesiac, as if he had worked to become passive and vague, whereas we know from experience that work clarifies ideas. If, in fact, we see the backward film as backward, and if the artifice of projection is not sufficient to reverse the sense of time for us, it is again because we are psychologically working by attentively assisting in its projection.

Causal efficacy cannot be directly apprehended in the mechanical collision of two balls transferring motion. On this point, Hume's critique is valid. But it is altogether different for meaningful activity, either within me or outside of me. I "am" my own meaningful activity. To say that I intuit it would even be insufficient, for my conscious activity is nothing other than "I." As for my instinctive and unconscious activity, I sometimes observe it from outside, discovering its meaning afterward, as if it were the activity of another. Even in this case, I grasp its meaning much less by induction than by intuition (except in those exceptional cases with which psychoanalysis deals). Instinctive activity, and especially the fully conscious activity of others, can be the object of intuition for me, just as my own activity can be, when it is not pure spiritual activity. It is certainly not by induction that I interpret the gestures of a man clinging to a dangerous slope as "effort not to slip." Or else one would have to admit that it is also by induction that I understand my own semi-instinctive gestures not to fall, as having the sense of "trying not to fall." The inevitable awkwardness of the sentence shows the awkwardness of the interpretation.[11] The past-future, in my own work-activity as well as in the work-activity observed in others, is thus a primary intuition. Hume's critique is valid, but it has an extremely limited scope. It applies only to theoretical cases where, precisely as we have seen, the reversal of time, realized by a film projected backward, is equivalent to a simple change of spatial reference points, and where reason is no more shocked to see ball *A* motionless, then set in motion by ball *B*, than to see ball *B* motionless, then set in motion by ball *A*. As soon as there is organizing action, or disorganization, it no longer applies. The sense of actualization, the intuition of its efficacy, and the past-future are all one and the same.

ENVELOPED TIME AND ENVELOPING TIME

There is one last point to clarify before drawing conclusions from this discussion. Psychological action, which organizes and informs, is the source of the past-future. But while psychological action taken separately and individually would indeed produce a past-future, it would not produce the *continuous*, seemingly impersonal past-future of everyday experience. A performed action has a past and consequently a future, but a future that has no reason to be regularly followed by another future. The time of the action is just an isolated impulse. The past-future of an action is a kind of pure quality, a raw material that is indispensable and fundamental to the past-future of daily time, but which alone would be as far from daily time as quantum physics is from ordinary physics—and, moreover, for similar reasons.

It is thus easy to understand why, when a psychologist who is determined to retrieve the immediate data of consciousness, or a phenomenologist driven by the will to set aside explanations and theories in order to return to the things themselves, confronts lived time, this psychologist or phenomenologist does not find linear, dimensional time. There is no continuous flow, no instants or intervals regularly ordered like points on a line, no empty time with an already-prepared past-future in which conscious action would take place, as in space. The psychologist or phenomenologist is perfectly right: actual experience coincides with the present but should not be confused with the abstract notion of the present moment. Actualization is not a linear past → present → future. For actualization is real existence itself, while the past → present → future line is only a schematization.[12] But the phenomenologist goes too far if he believes he can eliminate all past-future from his descriptions. He finds them under synonyms. Husserl's "retention" and "protention," Gaston Berger's "what comes from . . . " and "expectation," are more concrete and indeed better synonyms for past and future, but they are synonyms nonetheless. Just as the physics of elementary actions must find a way to join macroscopic physics, the time of psychological action must be able to join everyday time. It does so through the general process of framing auxiliary mechanisms, or channeled crowds, which we have studied at length:

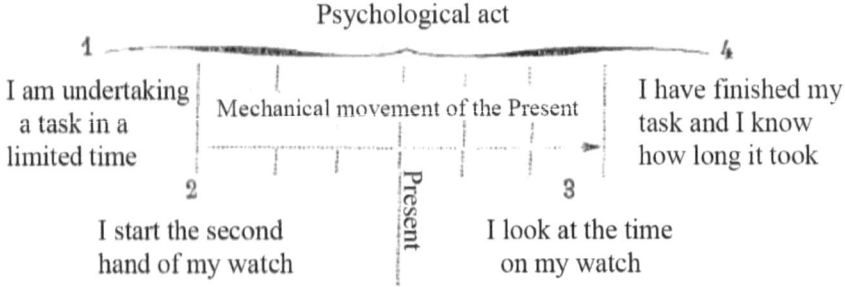

Figure 7.3.

The movement of a clock mechanism or the flow of sand in an hourglass serve as the auxiliary automation for all activities. The psychological activity encompasses everything, but its present moment, accompanied by the physical and statistical present moment of sand flow or the regulated release of a spring through an escapement, becomes a present that moves along a symbolic line. Its qualitative and primary past-future becomes a linear past-future. In the case of time as well, we could, in theory, do without auxiliary mechanisms and their assemblies [*montages*], through a psycho-physiological set-up [*montage*] or set*: the very one, for example, that allows for "trace reflexes" and serves for the approximate psychological estimation of elapsed time. The interposed automaton, or even the semi-independent play of the psycho-physiological set,* can only reflect the time of the psychological act by specifying it, just as a weighing scale reflects the will to choose based on weight and precision, without having the ability to choose on its own. At the bottom of auxiliary mechanical functions, there would undoubtedly be authentic actions, fundamentally of the same kind as psychological activity. There are quantum interactions in every expenditure of energy.

But these elementary actions only intervene in time-indicating mechanisms statistically, and through a blind game of step-by-step pushes, while encompassing and surveying psychological activity is unitary and purposeful. It takes advantage of the linearization of mechanical time; it uses it as a means of precision; but it does not owe to it the inherent past-future of all actualization.

Physicists have always somewhat mocked the philosophical subtleties concerning the nature of time, because they know very well that the philosopher himself will have to consult his watch. Eddington, for example, imagines a discussion between Bergson and the Astronomer Royal: "I rather think that the philosopher would have had the best of the verbal argument. After showing that the Astronomer Royal's idea of time was quite nonsensical, Professor Bergson would probably end the discussion by looking at his watch and rushing off to catch a train which was starting by the Astronomer Royal's time."[13] Similarly, Berger's paper on the mythical nature of time, recorded in the *Bulletin de la Société française de philosophie*, begins with this sentence: "The session is open at 4:30 p.m.," and the speaker apologizes, with fully conscious irony, "for having to stay within the limits of the time allotted to him," to prove to his listeners that time is a myth.[14]

However, it is the philosophical analyses that are ultimately correct. The Astronomer Royal's time, or the time of the physicist of thermodynamics, is only an auxiliary time, and it is as illegitimate to consider it as fundamental as it is to consider automata as providing the true model of living beings. The man who consults his watch seems to recognize the primacy of mechanical time, but this is not the case. Man has only made timepieces and clocks to serve as auxiliaries to his intuitive time. Clocks, like all machines, are only the extension of psycho-physiological organs or sets* and are nothing without the living organism. The psycho-biological actualization, the primary past-future, is not a simple functioning in time; it makes time. It involves creative information, which the present of the functioning can only mimic. That is why there is an absolute sense of time, even when the framed mechanisms are mechanically reversible. Two men conversing on the phone do not reverse the direction of time each time one of them listens after speaking and becomes passive after being active. Beyond the physical communication carried out along the line, there is creative actualization, both when an interlocutor listens and when he speaks, because he must realize the meaning. The man who consults his watch is like the man who takes a night train in which he will sleep until it arrives: he consciously and voluntarily subordinates himself temporarily to a mechanism that he has created. If my alarm clock wakes me up in the morning, it is because in the evening I wanted to be woken up at a certain time, wound up the mechanism, and adjusted the alarm hand accordingly.

Time in general—as opposed to the mechanisms that measure it—is not at the disposal of man. My consciousness frames the time of my watch, which I bought to know the time, and which other men have made. But my psycho-biological organism, capable of unifying a multitude of auxiliary functions and "making time" in its own domain, is also in turn dominated by centers of activity larger than its own, for which the lives of individuals must represent a kind of statistical flow similar to what cell metabolism is for the dominant "I." The speaker who chooses to subordinate himself to clock time in order to give a lecture on the mythical nature of time, or the man who sets his alarm for seven o'clock in order to be awakened the next day, cannot choose not to age and die, and not to be themselves, in their organic and psychological life, framed by the generations that preceded them and will follow them. There is nothing new or unexpected about this, since, as we have indicated, the framing pattern can exist at multiple levels, going up to the absolute Frame, passing through the living Species, whose individuals are in a sense only subordinate organs or cells. These multiple levels of Agents or centers of activity are naturally constituents of concrete moments in time, actual moments, and more or less extensive qualitative past-future times. We survive our cells, just as the human species survives us. Finally, the absolute Frame is not in time, although it is the ultimate source of all actualization.[15]

Chapter 8

The Mixed Origin of Information

The study of information machines gives us the positive certainty that information comes from a source that is trans-mechanical and, in the etymological sense of the word, meta-physical. But throughout this work we haven't been able to avoid doing a lot of metaphysics, not only in the etymological sense, but in the ordinary sense of the word. And the thesis of trans-mechanical information encounters a very serious objection, the same one that can be made against any kind of Platonism, against any explanation that appeals to a world of essences, values, or potentials: it would only be a pseudo-explanation by a useless doubling. If information of the actual world, the supply of information machines, is explained by the participation of individuals in some ideas or forms of a trans-actual world, what is it that informs this trans-actual world itself? What would we gain by just displacing the problem?

THE DILEMMA

We know with what vigor Émile Meyerson has emphasized this dilemma: either we understand a phenomenon by reducing it to a pure identity, by deducing from other phenomena the elements of novelty that it seems to contain; or we admit that novelty is absolute, and give up on understanding. We also know how Meyerson refused to allow attempts to conceal the dilemma with a pseudo "third solution": a recourse to the "state of potentiality," with which we would fabricate a semblance of explanation where identity is clearly missing.[1] The conception of information and the origin of information that we have opposed to that of cybernetics seems, at first sight, to be a pseudo-solution of this type. The man who composes a message or improvises an action; the organism that develops through self-structuration; or the beings that assemble themselves cannot be explained by the functioning of assemblies or feedback mechanisms. But we maintain that they are regulated by means of axiological feedback passing through a trans-spatial world. As a strict disciple of Meyerson might object, what difference is there between preformation and epigenesis if epigenesis is conceived as a preformation in which the form exists in advance, not in space, but in a trans-spatial world?

SOLUTIONS TO THE DILEMMA

Despite appearances, there are very few philosophical theories of the origin of information, because when faced with this problem the human mind always goes around in the same circle. We can distinguish the following:

1. Idealist theories of the Platonic type, according to which information in the sensible world is due to some ideal forms that descend into or manifest themselves in one way or another in our world. Despite the different aspects of their philosophies, Plato, Aristotle, and Leibniz adopt this schema.
2. Mechanist theories that reject the idea of a reservoir of transcendent forms and derive information from purely fortuitous combinations of elements existing in our space. Or the "geometric" theories in which the modes derive from a geometric necessity inherent in the nature of things, as in the philosophies of Descartes and Spinoza.
3. Dialectical theories, in which new forms appear according to a logical necessity immanent to the unique Reality, which spontaneously posits its various moments.

It is not very difficult to see that these three types of theories, although seemingly opposed, are in fact similar. These theories sacrifice novelty to rationality and resort to the notion of potentiality. This is obvious for the theory of substantial forms derived from Plato. It is hardly less obvious for the dialectical theories, and Meyerson was quick to emphasize it. Hegel, in his *Philosophy of History*, speaks of the "Spirit whose nature is always one and the same, but which unfolds this its one nature in the phenomena of the World's existence."[2] ["Spirit *knows itself*. It involves an appreciation of its own nature ... Universal History ... is the exhibition of Spirit in the process of working out the knowledge of that which it is potentially. And as the germ bears in itself the whole nature of the tree ... so do the first traces of Spirit virtually contain the whole of that History."[3] "[T]he Concept remains at home with itself in the course of its process," Hegel writes elsewhere, "and ... the process does not posit anything new as regards content, but only brings forth an alteration of form."[4]] It is less apparent, but no less certain, for theories of the Democritean type, as Octave Hamelin has rightly emphasized. Matter takes on, as Descartes says, "all possible forms." The possible then serves as the guide to the information of the world. The Spinozist modes are contained virtually in reality.

These three theories adopt the first branch of the Meyersonian dilemma. They want to explain everything, even at the cost of a "reduction" of novelty. Other theories, generally more recent than the first ones, adopt the second branch of the dilemma: they renounce the attempt to explain everything and allow absolute novelty at the cost of rationality. We can distinguish (1) Theories of pure experience, with pluralism, the "tychism" of William James or C.S. Pierce, and in general, absolute empiricism, which accepts the experience of novelty just like all other experiences. (2) Theories of emergence (C. Lloyd Morgan, Samuel Alexander), contingency (Émile Boutroux), creative evolution (Bergson), and pure freedom, that is, freedom without norms, values, or controlling essences, as understood by existentialism.

It is even easier to see the close kinship of these two last types of theories: they reject the notion of potentiality; they distrust concepts and logic, necessity, or determinism. While the three theories of the first type are theories of "metaphysical fullness," the last two believe in "metaphysical emptiness," which allows individual and unpredictable adventures, chances in the strong sense of the word (whereas chance in the Democritean and mechanist sense of the word seems to be only completing cases laid out in advance by possibilities).

Our own study of the origin of information through cybernetics has clearly led us toward theories of the first, Platonic type. How can we avoid the serious objections that it is so easy to put to this type of theory—especially that of "useless doubling"—and that our contemporaries, who are in general more seduced by novelty at the cost of rationality than by rationality at the cost of novelty, continue to emphasize?

When it is a matter of mnemonic information, of the repetition of a memory or of an organic form, the anti-Platonic objection is not serious. Since the billions of individuals who are born reproduce a specific form, the information of each individual must not be truly improvised and created each time *ex nihilo*. And if experiment demonstrates that there is no preformation in our space, we must conclude that there is a guiding preform beyond our space. A disciple of Meyerson would certainly be jubilant in the face of what would seem to him to be a simple naïve displacement of the problem. But he would be wrong. A re-production does not have to be explained as a pro-duction. Since the form is not original, it is normal to seek its origin elsewhere. Certainly, the idea of a trans-spatial type is at least as obscure as the spatial form. But what can we do about it? If the biological facts of reproductive information indicate that there is a super-nature beyond spatiotemporal nature (which, by the way, is not at all supernatural in the theological sense of the word), we hardly see why we would not follow these indications just for the futile reason that the hypothesis of this trans-spatial world does not give a total and "anhypothetical" explanation. The Meyersonian objection draws on a conception of science and philosophy that is too ambitious. Knowledge would never have progressed if we had always been discouraged in advance from explaining a by b because next it would be necessary to explain b by c, and so on. It is interesting to widen the sphere of what is known, even if we do not reach the point where the known becomes the intelligible. It is interesting to discover this world of mnemonic types beyond our physical space, to discover that there are several continents in "nature," taken in a wider sense than that of spatiotemporal nature. Instead of discussing the legitimacy and meaning of a metaphysics beyond science in absolute terms, it would be good to define a more modest sort of metaphysics, which would be a cosmology of the knowable, beyond the observable.

The question changes aspect when it becomes a matter of truly novel information, that is, of invention, and not of mnemonic information. Not that it is permitted to make a clear-cut distinction between memory and invention, as we have often seen. On the one hand, a psycho-organic reproduction is never standardized; it is always able to be fine-tuned. On the other hand, an invention is never free, it resembles a typical reproduction in being guided by the watermark of a possibility. There are some genres and species in technical inventions, just as in organisms, and often the same ideas are born in several brains at the same time, just as multiple individuals of the same species can develop thousands of miles apart. However, the difference remains significant, and a theory of reminiscence for inventive information is somewhat naive and abrasive. Consider for example more recent technical machines. It would be ridiculous to say that the inventor of the bicycle was guided, like the young slave in the *Meno*, by the intuition-memory of an ideal Type, controlling his efforts like a radar controls a D.C.A. automatic anti-aircraft gun. It would be ridiculous to say that as long as the bicycle had two irregular wheels and no freewheel, the inventors' minds were not quite satisfied—just as a man who is trying to remember the name *Warburton* is not satisfied as long as he can only manage to grasp the distorted name *Walliston*—but once the ideal mnemonic type is reached, the satisfying coincidence makes everyone happy and the model of the bicycle no longer evolves.

It is not enough to abandon the mythic clothing of the Platonic theory to render it viable. It seems that much more needs to be abandoned; almost everything in fact. Organisms themselves resemble artifacts in many ways, and while it is ridiculous to imagine that the idea of the bicycle or the camera was sitting in the ideal world for all eternity, waiting to inspire inventors, it is almost as absurd to imagine that the eye or the system of muscular levers of vertebrates ideally preexisted real organisms. That a trans-spatial world is mnemonically "conservative" of organic types or psychological ideas once they are invented is an almost inevitable hypothesis. But whether it is the place of Types explaining the very invention of organic forms during evolution, or the place of essences explaining the invention of ideas in individual consciousnesses, the hypothesis inevitably seems at the same time fantastic and naive.

But when we have abandoned "almost everything" in Platonism, when we have given up on believing that we can explain the information of our world by preexisting forms in a transcendent world, we nevertheless realize that something essential must be kept: namely, a certain dualism between values or meanings glimpsed by the agent, and the laws of the physical world that this agent channels in the direction of this value or that meaning. Even Bergson, who was quick to denounce the "ready-made" in explanations of novelty, retains a dualism that is fundamentally Platonic when, in a rather clumsy metaphor, he speaks of a "current of consciousness launched through matter."[5] Matter is only a vague notion and consciousness is not a current. It is what surveys and frames the auxiliary mechanisms that it assembles and arranges in order to force them to function according to its own axiological direction. But there is certainly, in all information, a sort of encounter between a conscious theme and harnessed physical laws. Let's leave the domain of abstraction and consider two examples, one very simple, the other very complicated.

a. *The teapot*—The teapot did not arise in the mind of its inventor from an already-formed ideal theme, but rather from a reasonable intention: to keep a liquid hot and to be able to pour it conveniently. This intention itself is partly based on knowledge acquired through contact with physical laws. However, it is still a relatively unformed theme. The determined form of the real teapot is born from the encounter between the intention and geometrical, mechanical, and physical laws. It first needs a central body, with an opening at the top, since a liquid subjected to gravity tends to spread out. The base must be large enough to provide stability. Moreover, it needs a spout or pouring channel. According to the law of communicating vessels, the top of the channel must reach the same height as the body. If a manufacturer were to design a teapot according to the model in figure 8.1—such as are used in intelligence tests—he would quickly perceive that the laws of physics protest. Without waiting for this to happen, he makes the pouring tip "conform" to the requirements of hydrostatics through a mental experiment based on knowledge of physical laws as much as by a properly intelligent act. In more sophisticated models, the handle must have a low heat conductivity so the hand that holds it is not burned; the lid must be stoppered by an interior rim, and so on. The form and dimensions of the teapot must also conform to the form and dimensions of the hand and the human organism. But this organism itself, we have seen, is born from the encounter of primitive vital "intentions" with physical laws, so there is no vicious circle in our analysis. Furthermore, the form of the teapot, like that of all household utensils, is standardized in a craft or industrial culture; it is preserved in a tradition or social memory, just as organic forms are standardized in biological memory. The history of technics can trace the evolution of these forms across different cultures.

b. *Astronautics*—Astronautics is a very appropriate example precisely because it involves the accumulation of the most sophisticated technics and even requires technics that are not yet fully developed. Astronautics is still largely in the state of mental experimentation. It thus allows us to observe the "becoming" of invention and information, because it is also in the process of "realization." Real rockets are already built. The effects of high and low gravity have already been tested on animals. We have begun to master nuclear energy, the only energy source capable of effectively propelling the vehicles. On the other hand, celestial mechanics is able to precisely trace in advance, from possible trajectories, the trajectories advantageous for a given purpose. But it's a faith, an intention of a spiritual and trans-physical order, that unifies all these partial technics, in order to aim at their realization. Human beings know in advance in what ways these technics are still insufficient. They identify in advance the incomplete areas that oblige them to further work and invention. Before voyaging into interplanetary space, they still need to make progress in the "space" of possible technics.

Let us consider more precisely the form the spaceship will take. This form is still, in many ways, a rather vague possibility, as vague as the form of the airplane was for the young Louis Blériot. An artist illustrating a science fiction novel wouldn't hesitate "to move from schema to image" and represent an interplanetary rocket imprecisely, using details

Figure 8.1.

inspired by V2s and stratospheric airplanes. But an engineer is already able to define the thematic possibility of the apparatus more seriously; for example, by taking into account the "mass ratio" between the fuel and the rest of the apparatus, by determining the necessary density of the protective plating, or by calculating the general form of the interior fittings. With a hand-held slide rule, he can eliminate in advance the "false" forms suggested by the mere association of ideas. In modeling the form of a new engine, what directs him is obviously his knowledge of the laws of physics, and not only "theoretical" laws, but also, in the Cournotian sense of the word, "cosmological" situations such as those prevailing in the regions concerned by human intention.

On the ground, Auguste Piccard's stratosphere balloon had, for eyes accustomed to ordinary spheres, the surprising form of an elongated pear, with most of the envelope hanging down to a sealed carrier, which was also very different from the usual basket. This unusual form was directly governed by knowledge of the fact that at an altitude of 20 kilometers, the gas would expand significantly due to the decrease of pressure. In the same way, the deltoid wing of modern airplanes is governed by the special aerodynamics of supersonic speeds. And again, we can easily imagine how everything in the spaceship must be governed by knowledge of physical laws. A teapot, in a spaceship, would be useless, at least if the journey were not made under constant acceleration. In a weightless environment, a sealed flask would be needed, since the liquid would not pour. [On the other hand, one of the important, and quite unexpected, parts of the spaceship would be the equipment designed to absorb the giant flash that would erupt on arrival at the planet visited, between this planet and the ship, the latter having been raised to a very high electrical potential due to the ejection of electrically charged particles.]

At the same time, we understand the indispensable double role of consciousness. It is, on the one hand, an intention and a faith [*foi*], prolonging the intention and faith that animate all living organisms. Human beings will take over the planets of the solar system, driven by the same faith that animates the lichen taking over an old wall. It is on the other hand absolute survey, that is, unified knowledge of domains where statistical physical laws reign. Consequently, it is able to anticipate and channel the inevitable effects of these laws instead of merely being subjected to them. A "possibility of intention" does not materialize on its own like a "possibility of functioning." Suppose that the astronauts would not have thought to neutralize the difference of potential produced by the functioning of the nuclear engine. This oversight would not put a stop to the physical laws at play; these physical laws would destroy the vehicle on arrival. In the manufacturing of basic tools, the method of trial and error can replace, to a small extent, the consciousness capable of mental experiment. But, in advanced technics, this method is impractical. The invention of the form must take place with foresight, by means of protections or preventative channels.

Moreover, it can be said that the invented form is initially nothing more than the set of preventive protections and channels that directly express the difference between an unsurveyed domain, where physical laws operate through step-by-step interaction, and this surveyed domain. Protective membranes and channeling tubes, the rudimentary organic or technical machines, are hardly more than that. An elementary organism is made of membranes and tubes or vacuoles. "To inform" a domain is first to enclose and channel the dangerous exterior forces. A further step consists of using the channeled physical laws by making them work according to the guiding intention. A combustion engine or a muscle motor channels the work of chemical energy. The interplanetary rocket itself is still, theoretically, only a tube opened at one end, and accelerated by the energy of the forces that it channels. The realization, of course, is terribly complex. The V2 rocket, whose principle is so simple, had such a complex piping system that it was nicknamed the "plumber's nightmare." The complexity of the spaceship will certainly be dizzying. The machine will be to the V2 what a mammalian organism is to an earthworm. But the fundamental principle of the information of a physical domain remains the same: a channeling of physical forces. Bacon's axiom "*Natura non nisi parendo vincitur*"[6] is not only a profound statement, it is one of the most profound and fundamental ever to be pronounced. It emphasizes the inevitable dualism at the origin of all information, and it provides the secret, which was missed by the diverse theories we have enumerated, for the production of a new form.

In the case of astronautics and the spaceship, every one of these theories appears inadequate to the point of being comical. The eternal idea of the spaceship is even more risible than the eternal idea of the bicycle. Equally risible is the mechanist interpretation that explains the production of the spacecraft either by mechanical sorting and natural selection, or by some sort of pure deduction, when the role of faith is as evident as that of mental experiment. As for dialectical explanations, one already feels nauseated just imagining the sentences that a contemporary Hegel would write on the subject. Let us move on to irrationalist theories. Engineers would shrug their shoulders at these ideas, when they fight against obstinate and inflexible physical laws, painfully seeking a narrow path between technical impossibilities that are harder and steeper than boulders. The astronautic adventure is nothing of an "adventure" in the existentialist and literary sense of the word. Human liberty, which will one day enable man to wander around the solar system and exploit all its resources as he sees fit, has little to do with the "metaphysical emergence" of an absolute freedom without norms. The slow and laborious manufacture of the necessary machines is nothing like a free emergence, nor a sort of spontaneous conversion of the intensive to the extensive, the schema to the image, through sheer mental effort.

AMBITIOUS CYBERNETICS AND EFFECTIVE CYBERNETICS

With respect to the Meyersonian dilemma, the cybernetics that adopts totalitarian mechanistic postulates, which we will call ambitious cybernetics, belongs to the first, rationalist group, among the monist mechanist theories. It has all the blatant insufficiencies. But cybernetics, as pure technics stripped of its pretentions, is instead an admirable illustration for understanding the mixed origin of information. Let us consider again the privileged case of astronautics. Albert Ducrocq has profoundly emphasized the particularly important role that questions of scale play.[7] A spacecraft, by virtue of the law of mass ratio, is inconceivable below a certain minimal dimension that probably represents several thousand tons. This necessary enormity will create serious implementation difficulties. To illustrate his point Ducrocq uses a very telling comparison. Suppose that organic evolution has resulted not in an intelligent vertebrate such as the human but some intelligent ants, who would have built a civilization in a limited corner of the earth, for example in Cornwall. These ants, with the help of machines of their size, or of the same scale, would have explored all of England. Then, with the aid of telescopes they would have been able to discover, in the distance, the neighboring islands, and even the coasts of the continent. But to cross the sea, they would need to build boats that are big enough to withstand the waves—gigantic compared to their own dimensions—and equipped with equally gigantic engines. The steering, like the construction of these boats and their motors, would only have been possible for them through stages of relays and servo-motors. A miniature instrument panel could only have directly commanded automated systems, which alone would have been able to operate the gigantic engines. This image is more than just an image, because man, faced with atomic energy and the plan to focus on an interplanetary rocket, is in exactly the same situation as an ant faced with the combustion engine and the boat to cross the seas. In both cases, the development of cybernetics and automata is an essential condition of success. Calculating machines, servomechanisms, and feedback automata are indispensable for allowing humanity to continue its conquest of the world.

The organisms themselves, at least those larger than viruses, were only possible through automations in relay. An adult man, relative to an initial cell, is even more gigantic than a 10,000-ton spaceship relative to an adult man.

The entire macroscopic physiology of multicellular organisms represents a group of power machines controlled by the intermediary of information machines. Organic forms are already the result of a mix of faith or axiological intentions, combined with the technical channeling of macroscopic physical laws. It is therefore not surprising that physiologists have so much to borrow from mechanistic cybernetics. But it is no less evident that they are wrong, if they believe they can cross the line and dogmatize about these mechanical relays. If the mechanist theories of the invention of forms are absurd with respect to astronautics, they are also absurd, and for exactly the same reason, with respect to the physiology of organisms. Effective cybernetics, detached from ambitious cybernetics, then appears as what it really is: an auxiliary of life and conscious intention, indissociable from life and consciousness.

REMAINING PROBLEMS

We see three main ones.

1. The Meyersonian objection is not philosophically resolved. It seems that we have not succeeded in puting both types of theories of novelty "on an equal footing." In fact, by linking inventive information to a mixture of intentional consciousness and adjustment to the laws of physics, we fall into the first type of theory: novelty was contained virtually in consciousness, the laws of physics, and the meaning and values aimed at by consciousness. [We have therefore arrived at a theory which is not very different from that of the *Timaeus*.] We have emphasized that the connections of passive assemblies were given in advance, as conscious connections of the active assembly; that there was, in a sense, *more* in consciousness than in the substituted automation; and that the machine was only an *extract* from the *unitas multiplex* that is the field of consciousness, where all possible connections are virtually present.
2. Any dualism or philosophical pluralism is unsatisfactory. To allow several principles or distinct domains to coexist is to leave things incompletely understood. But we have allowed the opposition of the world of meaning and of value, and the spatiotemporal world, to persist. We have also allowed to persist the opposition between psycho-biological individuality and its guiding ideal.
3. Finally, we have seen that inventive information in the atomic and microphysical domain is necessarily of a type very different to information at the level of "secondary" laws and macroscopic phenomena. The *"natura non nisi parendo vincitur"* cannot be applied in the same way to the manner in which a virus-molecule finds a way to repair and

reproduce itself, by capturing more simple molecules, and to the manner in which the animal uses water or air for respiration by channeling it through the gills or the lungs, or to the way that man uses a current of air to power a windmill, or an ejection of particles to propel a rocket. There are no real tubes or membranes in a virus, which is hardly more than a molecule.

Without claiming to escape these difficulties, which are closely connected, we can note that they largely neutralize each other. The spatiotemporal world, where the macroscopic laws of physics reign, is not a type of matter like the primordial and adversarial Matter of the old Gnostic dualism; it simply results from the multiplicity of individuals, which are truly primary. The spatial world is only matter in so far as it is colonized or colonizable by those individuals who are more enterprising than others. Connections, linked arrangements, channels, and machines only appear when an enterprising colonizer, whose directly governed empire has already reached a significant order of magnitude, deals *en masse* with the subordinate crowd of individuals, instead of annexing them in detail as it does at the level of microphysics. Respiration or assimilation in a large plant or animal, for which air and water appear as material, a continuous current to be utilized, naturally prolong the molecular respiration or assimilation, in which air and water play a role as ancillary molecules in a fully structured system.

With matter only appearing with the progress of colonization and as an obstacle or an auxiliary treated *en masse,* the only remaining philosophical problem is the duality of agent → ideal. But this duality is also created by the multiplicity of individuals. If there were only a single, simple Agent, facing the ideal world, there could be no proper psychological consciousness; it would be as if absorbed into this ideal world, in the way that, as the theologians say, divine will is one with divine understanding. Psychological consciousness as a *field* of consciousness—and not as a pure actualizing intention—can also only be conceived as a secondary phenomenon. The typical field of consciousness is the visual field that, when I devise some information—when I write a letter or draw—appears to me first as a blank page. In order to have this field at my disposal, it is necessary that I have eyes, formed of numerous sensory cells. The psychological expanse "upon" which I would draw the lines of my invention, controlling myself according to the ideal envisioned, corresponds to the physical world, which itself requires a multiplicity of individuals.

Yet, it is in consciousness as a field that possibilities and possible connections appear to be inscribed in advance. It is in relation to this consciousness-field that the substituted connections of the machine appear to be only a simple extract from the *unitas multiplex*. Consciousness is only "multiplex" secondarily to the colonization carried out by the original kind of "unitas" conscious intention. Visual consciousness, as a blank page for invention and information, is already a mix of trans-spatial and spatial. The field of consciousness as psychological expanse, corresponding to the space of the physical domain "surveyed," is the result of a crossover between the non-physical "dimension"—where the "I" aims for the ideal—and the world of crowds and multitudes, which are to be informed and connected according to this ideal, but also according to their own functioning. This is why invention seems to be virtual, and why the Meyersonian objection seems to retain its force. But it is necessary to remark that this "virtuality" is very different to the naïve "virtuality" that hides, as in a box, forms that would later emerge to appear in space. The "absolute survey" of the blank page gives everything and gives nothing. It allows for invention but leaves everything to be invented. The meaning or themes that I glimpse in a strange "nowhere," or in an imaginary dimension beyond the blank page, only roughly guide my pen in a way that seems frustrating to me, but it is precisely the place left for my real existence. The consciousness-field is only equipped with a "creativity" that is a virtuality of the invention of forms, and not a virtuality of forms. If I want to duplicate a square on a blank page, like the slave in Plato, I do not have any implicit reminiscence which could guide me. It is only through absolute survey that I can extend a straight line, trace auxiliary lines, notice some equivalents or inequivalents, and judge the evidence of failure or success, according to my intention. In short, I can take initiatives that are at the same time free and yet conform to the range of possibilities in which these initiatives unfold themselves.

Does the Meyersonian objection still apply to "creativity" itself? Is it just a word concealing the absence of a solution? It comes to the point where reason must give way to theology or metaphysics. Let's say then that this "creativity" of consciousness can only be conceived as a fragment of the cosmic[8] Creativity, of what we have, thanks to cybernetics, defined as the absolute Framing, the thread that traverses all individual conscious activities, themselves framing psycho-physiological assemblies and auxiliary mechanisms.

Chapter 9

Summary and Conclusion
(To the First Edition)

This work may appear to be mainly critical and negative. We have, indeed, attempted to argue that the mechanist postulates of cybernetics are both logically and experimentally untenable. Taken literally, they lead to the absurdity of a "perpetual motion" of the third kind. And, on the other hand, they clash with the facts. These facts reveal that all information machines, as much as ordinary machines, are always framed by a conscious and meaningful activity. A machine is never more than an ensemble of auxiliary connections set up by that improviser of connections which is consciousness. Information, as the communication of meaning, is only a particular case of information as the creation of form. All authentic anti-chance derives from connections, and all connection derives from consciousness. Turning to the machine in order to dispel the mystery of anti-chance and of the origin of information is then contradictory. Organic and psychological information, the organized and meaningful epigenesis of structures, in memory and in invention, cannot be explained mechanically.

The failure of cybernetics to understand the origin of information and anti-chance, and the purely apparent and superficial character of the success of its mechanical models—when it claims to understand the perception of universals,* learning,* communication between individuals, and the sense of time—are proof of the fallacious character of its mechanist postulates.

But to critique these postulates is not to critique or diminish cybernetics itself. Cybernetics represents one of the most remarkable advances in contemporary technics, science, and philosophy. With the mechanist postulates abandoned, we are not faced with cybernetics, *minus* something. On the contrary, we have cybernetics *plus* a powerful procedure for exploring the problems of life and consciousness, and for understanding the mode of attachment of structures and meaning, of physical space and the trans-spatial. The study of feedback mechanisms leads to a definition of axiological feedback and regulation, which sets up and envelops the feedback mechanisms, and of the axiological "space" that envelops physical space. It allows the definition of the generalized dynamics of which Leibniz and Cournot dreamed, without having had the means of establishing it with sufficient precision.

Chapter 10

The Problems of Cybernetics in 1967

CYBERNETICS AND "INFORMATICS"

One might observe that the word "cybernetics" has been less used over the last ten years. This is certainly not because the object of cybernetics—the technics and the theory of information—has revealed itself to be illusory or lacking true unity. The physics, biology, psychology, and sociology of information, along with their adjacent techniques, and the general study of communications and languages, continue to develop. But this very development is such that the specialists feel less of a need to resort to a general title. The word "informatics" would perhaps be more suitable, but it also lost coherence due to excessive development. It designates, sometimes exclusively, the computer's techniques of processing large amounts of information. And it also continues, along with the word cybernetics, to designate the general theory of information.

Large computers are not cybernetic machines in the strict sense of the word. They are not purposive feedback machines. They present some "digested" information to the manager who desires to be informed. They play the role of a secretariat that the management asks, "Give me a report on the employees, based on their salaries or seniority." They allow for a well-informed decision that takes into account multiple factors that are difficult to consider mentally all at once. The computer can also replace mental experiments: the manager can try out on the computer what the effect of some such decision would be—especially if the computers are coupled with some analogue calculators, for operational research or "optimization" in general.

A surprising effect of computers is that the departments of the firm that has decided to make this expensive purchase must reorganize their whole business to accommodate the computer. They also sometimes find that the prefabricated circuits of the computer (hardware*) are partly useless, and even detrimental to this reorganization. They may even find out that if they had carried out this reorganization first, they could have done without the expense of the machine.[1]

The modest performance of large computers, even when coupled with analogue calculators, seem not to add up with the public's persisting belief that they have a role to play in major political decisions. At the Geneva summit in 1965, even highly educated people eagerly questioned the Polish American mathematician Stanislaw Ulam, who was supposed to be in the know, about the role of decision-making machines in the Vietnam War.

While evidently false, upon reflection this belief is not absurd. A head of state would make better decisions if there was a way for him to be perfectly informed, with or without "informatics." But no machine will ever go searching for truly essential information by itself, if we have not supplied it with its "input." Assuming it had only "digested" equipment reports, a computer could very well have answered, like the French minister of war to the chamber in 1870, that "no button was missing from the army's gaiters."

TOYS

The cybernetic toys, whose archetypes are Grey Walter's "tortoise" and Ashby's Homeostat, perfected by Dr Sauvan to become the "Multistat," seem a little passé today. Assemblies [*montages*] for specific purposes such as perception, reading, learning, and so forth are usually preferred.

The cyberneticians have finally realized that, in the industrial machines made purposive via feedback, the purposiveness is double, or comes in two stages.[2] First there is the "production" of the machine itself, which is its primary end, its main line of material flow. And there is on the other hand the flow of information in service of the line of production, which operates in reverse to it. A boiler is designed to heat up, with or without a thermostat. *The feedback does not then "give purpose" to ["finalize"] the machine properly speaking*, as the cyberneticians repeatedly said. Every industrial machine has already been given a purpose—by the conscious human being who invented it—in terms of its main line of production and its output. The added feedback guides and improves the productivity of the primary purpose. But without this primary purpose, there would be nothing to guide, improve, or control.

However, a mechanical toy, unlike an industrial machine, has no primary purpose [*finalité*]. It does not extend and serve the primary purpose of humans. The toy is useless—it is only meant to amuse, by its movements and pirouettes, without these movements being productive. Japanese industry, for example, has made a box that is a kind of "pure toy": when it is opened, an articulated hand comes out and closes the lid again; that is all. In the same way, we can make toys that are only feedback mechanisms. Their motive is only to feed on this feedback itself. A toy steam engine does not activate anything other than the safety valve and the regulator. In a simplified form Grey Walter's "tortoises" immediately have great success as toys. They do not do anything; but they do it "with care" and, it seems, "intelligently." Ashby's Homeostat and Dr Sauvan's Multistat do not do anything either. They are only feedback systems, connected and controlling each other. Without a primary purpose, their secondary purpose means nothing.

TEACHING MACHINES

Pedagogy is what has been most directly affected by cybernetics over the last several years. Cybernetic pedagogy has been the object of a real passion, especially in progressive circles.[3] Predictions about the future "knowledge industry" are sometimes quite delusional.

Teaching by machines, or programmed teaching, comes straight from "Skinner's box." But if Skinner is the father, Pavlov is the grandfather. In Skinner's box, designed for instrumental conditioning, the rat understands by itself that by pressing on a pedal, it will make a food pellet drop. The rats benefited from a scientific pedagogy before people did.

We will remark (1) that the rat, in the box, is very "demanding" (it is hungry). (2) [I]t is the only living and conscious being: it must find the pedal as a means of getting food by itself. (3) [T]he "instructor" (Skinner) has disappeared; his role is materialized in the arrangement of his box, and this role consists in presenting a well-defined problem to the well-behaved rat: to find the pedal and its function. (4) [T]he rat, after success, is not only "*informed that . . .* " there is a pedal. Instead, the rat is equipped with a *knowledge*, which is more permanent, and in all likelihood corresponds to a set-up in its nervous system: the pedal is something "to press" in order to get food.

To grasp the essentials of cybernetic pedagogy, it suffices to replace the rat by the student wanting to learn spelling, a foreign language, or basic algebra, and to replace the Skinner box with a machine presenting a programmed and progressive problem, and bringing the possibility of learning a succession of basic knowledge elements. For example, the student who learns English is (sensorially) "informed," by a series of pictures, and a series of words, seen and heard; he must then complete a phrase with a word missing: "A . . . (*dog, man, cat, book*) . . . on a rug," by a good choice of word-pedals, and he can only continue if he finds the right one. The reward, here, is the success itself.

Cybernetic pedagogy reverses the schema of so-called traditional pedagogy, in which the teacher is active and the student passive; in which the teacher speaks, transmitting information, while the student listens. This goes for the teaching of the Koran, the lecturers' courses at the universities, and also teaching over the radio—which proves, by the way, that it is not sufficient to have a machine, even one for "audiovisual" information, to be included in cybernetic pedagogy.

In cybernetic pedagogy, on the contrary, it is the student who plays the active role. He seeks the right way, the right tool. The teacher, or rather his programmed machine, only disciplines the effort: the student is informed of success (by feedback), and as a result (according to the psychological laws of learning), he memorizes the right way, and can in this way progressively establish in himself an "internal informational chain." The teacher is so passive that, after having spent a lot of time programming his teaching, once in class he can have the impression of wasting his time, because he does nothing more than supervise the activity of the students.

The student must be strongly motivated, not by an instinct, like an animal, but by an "ideal." It is the condition *sine qua non*, which is completely opposite to the mechanical toys made for "learning." He must invent the right solution at every moment, in a sort of vicious circle that is always overcome and in which he fills a "blank" by informing himself.[4] Even at the most basic level of learning a language or a manual job there is no training, but self-training, which is very different. (Although all training supposes a minimum of self-training, that is, of little initiatives by the student.) There is a controlled autodidacticism, a mental creation; a controlled, but spontaneous, mental embryology.

Despite its apparently mechanistic character, cybernetic pedagogy is the antithesis of mere training. But if it simplifies the student's work too much, it actually goes back to a quasi-training, as mechanical as the reciting of the Koran or Catechism in the old times.

In the education of young people and adults, there comes a moment when it is necessary to give them the time to succeed by themselves, not in a succession of tests coordinated by the instructor, but through the completion of their own "little work" ["*oeuvrette*"].

LEARNING

In theoretical terms, the most important point to grasp about cybernetic pedagogy is the difference between the reception of *occasional*, functional information, which guides or shifts behavior, and the reception of *structuring* information, which is not meant to modify the immediate behavior of the receiver, but to modify the receiver himself by becoming knowledge within him.[5] One can listen to a phrase in English; (1) in order to obey and respond to it; (2) in order to learn English. Through a desired vicious circle, analogous to the process, "Suppose the problem is resolved," the user of cybernetic pedagogy waits for the learner, by invention or chance, to respond to the occasional information as if he already possessed the corresponding code, or internal set-up. The experiment indeed demonstrates that it is a quick means of learning. When the learning is successful, the learner is no longer distinguishable from an organism reacting "instinctively," that is, according to its nature—as we understand it according to its spatial structure, or its spatial structure combined with some trans-spatial elements—on the occasion of information to be used immediately. One speaks one's maternal language, which has nevertheless been learned; one responds to a phrase with a phrase, without needing to "consult" one's memory or one's grammar book. "Integrated" knowledge is not an internal source of information, "observable" or consultable like a dictionary. The knowledge is "participated in."

If they are to remain within the mechanistic framework, the cybernetic theories of learning must then explain how a flux of information received by machine M can become a new structure in the transformed machine, M'. They must explain how a feedback or programmed machine can, in the succession of information received, and according to the results obtained, not only respond *according to* its assembly or program but modify them. Thus, the modified versions of IBM 704 and 709 have, to use Wiener's expression, a second-order program (*higher programming**); they are programmed to modify themselves in their basic behavior, in a certain manner, which depends on the results obtained the first time they run. MIT's chess-playing machine first has only the mechanisms [*organes*] that allow it to place one piece on the chess board, a temporary memory, a permanent memory, some information combiners, and some transmission mechanisms. When it moves a piece, the movement is recorded in its temporary memory. If it doesn't conform to the rules of the game, a human "partner" signals to the machine with a touch, and the machine transfers the "memory" of the movement made to its permanent memory, accompanied by the indication: "not allowed." Then, it will compare the initially decided movement with the disallowed movements before acting, and act accordingly. It then learns the code of the action; it puts it in its structure, without this mode having been preprogrammed in advance. In the same way, it learns not only the moves that are allowed but the ones that are effective.

This certainly appears to be a model conforming to learning, and not a simple "simulator." The double memory, temporary and permanent, for example, very probably has some real correspondences. Grey Walter has already used the double memory in the installation of Cora (a simulator of conditioning) that, added to his automaton, transforms it into a *Machina docilis*: a temporal overlapping of stimulus I with stimulus II is recorded and stored by memory 2, and preserved by a third memory, in the form of low frequency electrical oscillations.

In the chess-playing machine, it is the machine that plays the role of the active student. The human partner only plays the role of the teacher supervising the student's assimilation of the teaching program. Theoretically, the person could be replaced by another machine, II, which has already been "taught," and which communicates its integrated "knowledge" to the first machine, I.

On closer inspection, doubts arise. Machine I only appears to be active: its attempts are corrected from outside. The correct behavior is instilled in it progressively instead of being inscribed in its structure in advance. That's the key difference to a machine that doesn't learn. It is only capable of learning in a way that is externally "forced," whereas a child "can deduce, from what he has grasped from the external learning to which his surroundings subject him, its rule of conduct in the face of a new situation: he learns to learn, he practices self-learning."[6]

The analysis of the game of chess simplified to nine squares (Martin Gardner's *Hexapawn*[7]) where the moves are represented by the choice of a "matchbox" for a situation, and where the automaton, according to its very structure, would become unbeatable after having lost eleven games, has demonstrated that it would be impossible to make an automaton that could play real chess according to the same principle. This is because it would necessitate more "matchboxes" than there are atoms in the Milky Way, and a time just as astronomical to calculate the valid combinations.[8] Moreover, it wouldn't even play chess, properly speaking: the outcome of the game would be entirely predetermined, since every decision capable of leading the opponent of the automaton to victory would be disallowed from the outset.[9]

A true model of learning that would not be reduced to a simple *Machina labyrinthea*, with a rapidly gigantic number of combinations, would involve much more challenging implementations. Notably, it would require models for the perception of forms, for recognition and the establishment of categories for analogous forms, as well as the development of a "vocabulary" corresponding to these categories, and a syntax of use. As Michael Scriven remarks, it is a lot more difficult to create a controlled imprecision than a controlled precision, whereas the reverse is true for human beings.

THE PERCEPTION OF FORMS AND READING

"Learning machines" are like the "conquest of the cosmos." The progress realized so far, in virtue of its little significance on a truly cosmic scale, has mainly demonstrated that the goal is barely attainable. A voyage to the Sirius system or the Andromeda nebula is completely out of the question. The mere "perception," or "recognition of forms" (by advanced scanning techniques aided by simplification methods using key effects) is already challenging. It is only possible on condition that the forms to be perceived are sufficiently standardized. The recognition by the machine of the letter A as A depends on something more than the search for the coincidence of the form with an exemplary model [*cliché*] in its "memory." Some concept of "A-ness," if we can put it this way, must be formed.[10] This quasi-concept, being mechanical, can only be found in some traits common to the various forms that the same letter can take, while being able to be distinguished from the forms of other letters. To achieve this, we can proceed by first scanning the letter with a constant frame and recording the groups of characteristic matching points of the letter as a sequence of signals. Then we search for the maximal simplification, or the minimum of identification functions, by analyzing these groups of points solely from the perspective of their internal homogeneity or heterogeneity, for example. The machine itself is used for this search: we try all the possible "templates," and we select the sequences that are sufficiently discriminating and have an acceptable probability of error.

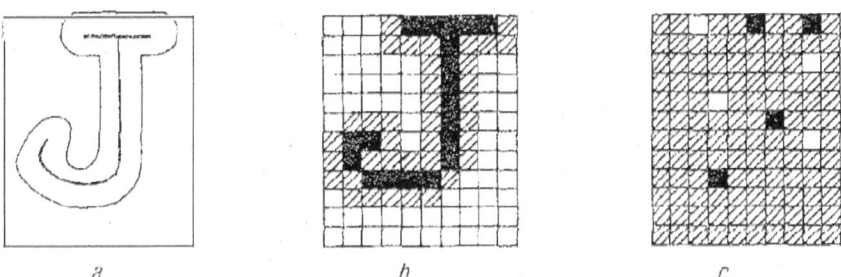

Figure 10.1. Minimalization of identification functions (after Deweze)

a. *original character and its envelope;*
b. *outline type (envelop-character quantified in 3 levels);*
c. *reduced outline type.*

An analogous procedure can be applied to other perceptual discriminations, in man as well as in machine. For example, R. N. Shepard has demonstrated that the auditory or visual discriminations of Morse code signals are essentially made on the basis of two characteristics of these signals: (1) their length and (2) their degree of internal heterogeneity. The human reader spontaneously employs methods similar to those of the machine. For example, we recognize, especially in handwriting, an *m*, an *n*, a double *n* (*nn*), a double *m* (*mm*), by the different length in the same internal homogeneity.

Of course, reading by a magnetic head or optical system is easy when the writing is completely standardized. The machine can then employ any coded external criteria. For instance, the American Bankers Association's* machine for automatically reading standardized magnetic characters. But clearly the real problem is the reading of letters and writing that is not standardized, about which the machine would have no *a priori* geometric or topological information. It would need to be programmed in such a way that it would devise its own criteria for identification.[11]

In theory, a reading machine can always succeed in devising some criteria for identification through "forced learning"—if one can even speak of learning in this context. Because the machine—or its various assemblies—is here no different to a series of tools that a man tries out in order to keep the best.

TRANSLATION MACHINES

The discouragement of researchers into automatic translation is notorious. The work continues, because the linguists and professors of language have realized that, in any case, it is a good method to act as though they are seeking to make a translation machine. This aim, feasible or not, is a good guide for the analysis of a language or of language in general. It is like cybernetic pedagogy or operational research: it quickly runs into some impossibilities and the dilemma of being either simplistic or immensely cumbersome (when breaking down questions in order to exhaust their complexity). But it is useful, because it obliges those who prepare the program to be precise in expressing their thoughts and to detail the problems instead of trusting their intuition. As Deweze very appropriately said—although this is true for the history of all technics—it is a question in all these cases of a "mutual teaching" between man and machine.

A translation is not a mere decoding. A machine encodes and decodes easily, because the transformation concerns not meaning, but simple equivalent supports of a meaning. Substituting *b* for *a*, *c* for *b*, *d* for *c*, and so on, to recover the original text, clearly does not present any difficulty. But substituting a French phrase for an English phrase with an equivalent meaning is another matter, because the correspondence is not, in general, of one word to another, but of one unit of meaning to another, by means of phonetic or graphic supports without a one-to-one correspondence. Consequently, it is impossible to materialize a transformation code that would conserve the meaning. From one text to another, everything takes place as if the translator must pass through the non-spatial region of meaning. And in fact, the human translator always mentally "realizes" the meaning of the word, of the phrase, of the translated text. The important point, emphasized by Louis Couffignal, is that up to the present day, the only mechanizable information (in terms of its conservation, transmission, and transformation) is unequivocal information: that is, information whose support corresponds to a single semantics. When this is not the case, we cannot take the shortcut of a decoding-translation.

An even more important point is that the difficulty of automatic translation confirms that it is impossible to define information by the simple measure of order or of anti-disorder (negentropy). This order is only the physical support of a semantics (that is, meaning, expressivity, and psychological effect in general). Information is the "semantics-support" ensemble. Yet the psychological effects of a support are capricious from the point of view of orthodox determinism: they are not without laws, but these laws are not expressible solely in terms of spatiotemporal structure. The structures of the support can only be isolated for their physical treatment (such as sending a telegram, recording, etc.) But the psychological effects depend on a *sui generis* combination, which can be compared to the perception of depth in a two-dimensional picture on a surface. We can trace and televise this picture, and thus remotely prompt the perception of depth in the viewer's mind. But this promoting, or guidance, is not infallible—as the study of perceptual illusions, where the psychological and even cultural context plays a significant role, proves.[12] To achieve or maintain the psychological effect of depth, or more generally, of meaning, it is often necessary to adjust the support.

In translation machines, exploring the context within the support is the only possibility for "guidance," the imperfect but only possible equivalent of the human translator's "ascension" to the region of meaning. This exploration is often effective, just as the exploration the observable context of the pictured object through the eyes is effective for perceiving depth.[13] Without the context, the human translator would hesitate as much as the machine.

But it is obvious that the context *of the support alone*, as spatiotemporally observable, is not the equivalent of the psychological context of the emitter. In order to translate into English the old French song

> *Orleans, Beaugency,*
> *Notre Dame de Clery,*
> *Vendôme, Vendôme*

in a way that would get a similar psychological effect, the exploration of the context would not help. It would be necessary to find the names of some English villages with an evocative power that would be similar for an English person (as Waterloo—this name of defeat curiously given to a place, Allais Alphonse said—is for an English person the equivalent of Austerlitz). In all these cases, the support is so closely linked to the psychological effect that they cannot be separated.

The similarity of the use of contextual clues by both machines and human beings greatly encourages advocates of automatic translation. Their reasoning is as follows: humans don't interpret an information support without clues, otherwise they risk being mistaken. Since clues are physical, they can also guide a machine. However, this similarity is partial and misleading, because there is a psychological "context," and not only a context of the support, in all emission and reception of information in the full sense of the word.[14]

"MENTALITY"

This word is a sort of scientific euphemism used to replace words such as "spirit," "consciousness," and "psyche." According to Dr. Jacques Sauvan, "the approach to mental functions is subject to an extremely severe taboo, of metaphysical inspiration." The taboo is not very effective, because since Grey Walter, Ducrocq, and Ashby, we have made such efforts in this direction that it would take too long to examine them, even briefly. Successful or not, these efforts have not led their authors to the gallows or the stake for being sacrilegious.

According to Dr. Sauvan, the Multistat, combined with his own Model S4 and Model S5, is capable of choosing not only means, like Ashby's Homeostat, but goals, by changing its "range of satisfaction"—for example, if it is subjected to an excessive influx of information. S4 is capable of what Dr. Sauvan calls an "informational homeostasis." It protects itself against disruptive information and welcomes "meaningful" information through some privileged pathways.

This would be a model of instinct, or more generally, an "epigenetic mechanism." Thus, Curt Richter's rats, or children suffering from an adrenal problem, "instinctively" choose their food according to the optimal amount of salt. They have no innate instinct in this regard; the choice is guided by an organic homeostasis, and there is no preformation. However, they must isolate a certain type of stimuli-information. Even embryogenesis, which appears to be strictly programmed as an automatic unfolding, is only self-adaptation to some rigorously filtered information. S5 is a model of memory. Its memory forms a linked network, where everything is interconnected (despite some subsections, with certain reciprocal inhibitions). In the same way, according to its creator, it has the power of creativity, imagination, intuition, association of ideas, and abstraction, through the choice of an optimum pathway in the linked network, which is explored in parallel. Dr. Sauvan remarks that this leads the automaton to a quasi-paranoiac "belief" is that its behavior, or its ideas, are not only the best, but the only ones possible. Dr. Sauvan concludes that this is precisely why our own brain, which functions according to this principle, cannot admit that the functioning of the machine is of the same type as its own, hence closing the loop.[15]

What then would the synthesis of S4 and S5 be capable of? The resulting system would be self-finalized, self-determined, and fully autonomous. It would even be capable of true reflection.

Instead of entering into a detailed discussion, it is interesting to emphasize that besides cybernetics, there are two major figures responsible for these tentative models of "mentality." First of all the movement of behaviorist psychology, following the thesis of Gilbert Ryle in defining the "mind" not as a sort of substance or specific reality, but as simply a name given to certain dispositions or functions of elements that anyone would consider as qualifying as "spiritual" or "mental." Here are some examples of the sort of adjective or noun to which the words "mental" or "mentality" belong. The Constitution of Britain is not a thing that we find beside Parliament or Downing Street. It is something that has a superior logical order. When we visit Parliament, Downing Street, or Buckingham Palace, the Constitution still remains to be seen, somewhere else. The adjective "mental" does not designate a mode of being, a status, such that we can reasonably ask, of a given thing or event, if it is mental or physical, in the mind or in the external world. "To speak of the mind of a person . . . is to speak of capacities, possibilities, inclinations, that this person possesses, to do and to undo certain sorts of things in the ordinary world."[16] It is in this sense that we are intelligent or lazy, exactly like a glass is "fragile," a clock is "precise," or a car is "fast."

We see the pleasure that such a conception can give to cyberneticians searching for "models" of "mentality." Let's apply it to various finalized machines, or to electronic machines for calculating logic or mathematics. They behave like conditionable, or even instinctive, animals; or they attain some good operational results. It is neither more nor less legitimate to speak of their "mentality" (of an order logically superior to that of physical phenomena, of the electrical currents passing through their valves and conductors, but not distinct, like a substance apart in a world of parts) than to speak of the mentality, intelligence, or laziness of a living human being (of an order also superior to, but not distinct from, the plane of nerve cells or nervous impulses). In this sense, a machine has a body—its visible mechanisms—and a "mind" ["*esprit*"], which is nothing other than its functional output, just like the Constitution of Britain, or a clock, or a car, or a glass is "effective," "precise," "fast," or "fragile." To understand a message is to react in a certain effective way to the message, or to correctly relate it with some "memories"; it is not to experience a kind of "mental sensation."

Alan Turing argued that if a robot could be constructed in such a way that it responds to an interrogation like a human being, we have to admit that it "experiences" something, or at least that we have neither more nor less reason to believe this than we do for a human being.[17] Since that time, scientists have had less interest in this question than philosophers, theologians, and artists, perhaps because they see best the difficulty of constructing a machine that gives even basic responses.

Here, for example, is a dialogue between a man and a machine, "Raphael":

Man—How many fingers does John have?

Machine—I don't know if a finger is a part of John.

Man—There is one hand on each arm.

Machine—I understand.

Man—There are two arms on each person.

Machine—I understand.

Man—How many fingers does John have?

Machine—How many fingers per hand?

Man—A hand has five fingers.

Machine—I understand.

Man—How many fingers does John have?

Machine—The answer is "ten."[18]

The dialogue is not very sophisticated. (We can nevertheless admire the good sense of the machine, which in contrast to a lot of human beings listens to the scientist before asserting anything.) But the reader understands that the machine's "I understand" has no more significance than would a transparent sign, attached by the vendor to a power socket to test electric lights, which would display the words: "Dear Customer, I am working properly" when the lamp is turned on.

In the face of this type of performance, we do not yet feel the need to found a society for the prevention of cruelty to robots, as Guido Calogero suggests,[19] or to ask, as Michael Scriven does, if it would be possible to construct a lying robot. According to Scriven, lying would be the proof of a difference, therefore a duality, between experience and expression, and therefore proof of the existence of "experience."[20]

"MENTALITY" AND THE ASSIMILATION FUNCTION

Louis Couffignal has drawn a parallel between the "function of mentality" and the biological function of the assimilation of chemical energy extracted from food: "We can say that the organs (the digestive tract, the respiratory and circulatory systems) have energy for their 'raw material,' which they extract from data constituted by air and food, that they supply to the other organs . . . which in turn give chemical energy in the form suitable to each cell."[21]

A critical remark here: François Bonsack has pointed out that it is necessary to correct the language commonly used.[22] It is not the energy of food that we consume, but only the usable energy, or negentropy. Not only do we not consume energy, but we must say that energy doesn't play any role, that it doesn't have any value by itself. It is not consumed (which would be incompatible with the principle of the conservation of energy); we find it quantitatively intact after having used it. What was consumed, destroyed, annihilated, after the assimilation, is negentropy, that is, the order or the non-disorder of energy.[23]

Now, Couffignal continues, it is another "raw material" that the organism receives from the environment, just like energy: information. All the informational operations—transmission, conservation, and combination of information—can be carried out by human beings.

This conception can be criticized. Indeed, recall that information is itself constituted of (1) a support and (2) a semantics—a collection of meanings, which is its essential element. The psychological meaning is not always a simple aspect of the support, as demonstrated by the difficulties of automatic translation. It has a certain autonomy. The support is essentially the language, transmissible as a series of signals (in general by the modulation of a carrier wave).

Yet, as Shannon and Wiener have demonstrated, the supports of information do indeed represent negentropy, or "ordered" energy. At the level of supports, the function of mentality is then quite similar to the biological function of the assimilation of food. A modulated carrier wave, received by the ear and "assimilated" by the auditory zone, is quite similar to the meat or sugar whose structured chemical energy is assimilated. But what about the level of meaning, of which the elements, once again, are not always in one-to-one correspondence with the structures of the support?

The cyberneticians have not attached sufficient importance to a paradox— signaled by Schaffroth and commented on by Costa de Beauregard—to which we have drawn attention in our own way in chapter 6 of this book. Suppose that the support of information is a set of forms in the physical sense, which can be taken in with a single glance—playing cards spread out, books on a library shelf, letters on the keyboard of a typewriter, and so on. As soon as the structure of the set is "perceived," that is, insofar as it *takes visual consciousness for a support* (and no longer the electromagnetic field, or the photograph, or the television screen), it no longer matters whether the elements of the set are in objective order or disorder, and represent more or less negentropy. Even if the cards and books are arranged haphazardly, "I" can recognize a particular card or book, without step-by-step optical scanning—and without the use of pins, perforations, or magnetic indicators. The psychological information of the visual field is independent of the order or disorder of the objects represented in this field. I can see my inkwell on the table, wherever it is.

The semantic assimilation of information, unlike the assimilation of food, is at least partly independent of negentropy and its support. I know where the ace of clubs is, or the English dictionary, or the letter N on the keyboard, even if the elements of the set to which they belong are *poorly* or haphazardly arranged—whereas my organism does not thrive if it has to assimilate only a "disordered" protein as food.

PATTERNS*[24]

The practiced typist improvises some schemas of taping, in coordinated "sets" on her keyboard (where the letters are not in alphabetical order). Even if she types what is dictated to her, she needs to transform this auditory information into spatial motor schemas, without visually scanning the keyboard. However, she can also compose directly on her keyboard, and in this case, she improvises the *schemas* of the words in her sentences before the motor schemas of the letters for typing each word. As linguists long have, the cyberneticians Grey Walter and Couffignal have perceived the importance of spatial or temporal schemas: chains and sequences of chains as means of access to the analysis of the semantic part of information.

The English word "pattern" is often used to designate these schemas. It comes from the French word *"patron,"* and it has roughly the meaning of "template for a dressmaker" (standard form for the cut of a dress). Unfortunately for the anti-Franglais purists, the word *"patron"* would too often be equivocal. The pattern is "anti-chance": a child hits the keys of a keyboard haphazardly, while the typist hits according to the motor "patron" of the words. Even if the child runs his finger over the line of letters, in the order on the keyboard, he produces nothing meaningful, despite this physical order. To produce some pattern, and therefore meaning, it is necessary to break this order.

The psychological effect produced by a physical action is generally perceived according to patterns, sometimes arbitrarily, such as when we introduce a rhythm into a regular succession of sounds or some imaginary figures into a randomly distributed set of points. The pattern is individualized in consciousness; it is stored in memory, it can be transposed to different materials (like a *patron* for a dress), and it can be recognized in a confused set. It can be decoded, designated, imitated, and completely modified. A billiards player does not mistake the ball—despite the two white ones being indistinguishable physically—because he makes, or sees, the pattern of the shot that he or his adversary is taking.

The meaningful pattern can be superimposed on the sensory pattern. It becomes so intimately united with it that it is artificial to separate the two patterns, as perceptual illusions demonstrate. But it is still the meaning that is active, and that activates the uniquely resulting pattern. This is particularly evident, for example, when we perceive meaningful figures in a Rorschach test plate, or in a pseudo-artwork, such as a poem, a piece of music, or a painting randomly generated by machinery. We believe ourselves to be reading or hearing passively, even though we are in fact active.

THE AUTONOMY OF "SEMANTEMES" AND THE THREE TYPES OF CONTINUITY

Unintentionally and despite its mechanist tendencies, cybernetics will have thus contributed to the introduction of the notion, which is widely accepted today in many fields, of the autonomy of meanings, of patterns-as-products-of-meanings, called "semantemes," or "semantic genidentities."[25] This notion is very revolutionary in relation to classical science, and the comparison is very illuminating.

Naïve physics (of ancient atomists, or of seventeenth-century mechanists) believed only in the reality and temporal continuity of matter. Then, energetic physics defined another type of reality and temporal continuity: energy. When ball A hits ball B, it's not just the physical balls that are conserved, it's the motion and kinetic energy that transfer from one physical "support" to another at the moment of impact. A wave is just as real as a piece of matter: it has a temporal continuity, a formal genidentity, just like its "support": the string or the molecules of water. Then, we perceived that energy—or the formal genidentities—could be primary and autonomous, and surpass material support. A field of electromagnetic or gravitational energy has no need of ether; it is a sort of matter-energy in space-time. George Gamow said that it is like a snake, which transports its wave movements with it. An atom called material is a system of waves, and it would be vain to search for a sub-matter as a support for these waves.

Yet patterns and semantemes represent a third space of continuity. A word, for example, does not owe its temporal consistency to its matter or its form. If it is passed from mouth to mouth or from dictionary to dictionary, this is not as a wave. It owes its consistency to its meaning, of which it is the carrier symbol, the formal support. The relation is quite free moreover, since words change meanings and meanings change words. The same as wave energy is most often tied to a material support, but can also be detached, a semanteme is tied to the form of an information-support, but it can also be detached. In translation, the (nonmechanical) translator reconstitutes the meaning, in order to pass from one support to another. In communication, which is the reverse, the emitter passes from meaning to a support, as the receiver passes from this support and its negentropy-information, to meaning. We search for words in order to express our thoughts. In the history of culture, what interests the historian are ideas, themes, schemas, or patterns of institutions or of techniques, and the like.

Thematic and semantic continuities, while resembling material and formal continuities, have a very peculiar character, at least at first sight: they seem to be independent of space and time; they seem to be able to enter and exit them. An idea can be lost, rediscovered, then lost again. This is either because its support remains without being understood and is then understood again, or because, if all support is lost, the idea is reinvented. One can pick up a problem again, or forget it or leave it aside, and take up something new. One can also resume a learning process in progress, after having done something else in the meantime.

Is this independence with respect to space only an appearance? We here touch on what is probably the essential division between two rival and irreducible interpretations, not only of cybernetics, but of philosophy and of science.

1. We can refuse to admit that meaning and semantemes are really detachable from space-time, and consider any such conception as mythological. It is then necessary to admit that the idea, the meaning, when it seems to be "lost," is pictured in the (material) brain, or in some auxiliary mechanical supports of the cerebral memory. This amounts to considering the semantic continuities as fundamentally illusory, by reducing them to the material or formal continuities.
2. We can frankly admit this autonomy with regard to space-time. This is what we have done by considering mechanical feedbacks and materialized "controls" as only limited and degenerate forms of ideal feedback

and control. Couffignal, who nevertheless remains "spatialist," is not very far from this conception when considering "mentality" as a sort of information assimilation function, since this comparison suggests that the semantemes are autonomous, and can be detached from their support.

It seems to us that science, since 1954, has itself made a number of steps in this direction—not only the human sciences, which have always postulated the autonomy of ideas, without always admitting to this postulate—but even the physical sciences. They are today a lot more attached to the conservative principles in space-time, even for matter-energy (which authorizes the same detachment *a fortiori* for semantemes). If energy is conserved, negentropy is not conserved. We have known it since Carnot and Boltzmann, and Bonsack has emphasized it again vigorously: Carnot's principle implies that something leaves the world. On the other hand, according to Fred Hoyle, we must admit a continual creation of matter, which thus appears in space (or creates supplementary space of some sort since the average density of the cosmos is constant, the expansion counterbalancing the creation).[26]

We do not then see what would be mythological about admitting for the semantemes what all physicists admit for negentropy and some for matter-energy. If there is some mythology, it is rather in the arbitrary postulate that supposes without any proof that all subsistence of meaning is always fundamentally material memory. This postulate is even more inadmissible since, for contemporary physics, the material persistence of a mechanical memory is itself, like all that is material, the persistence of a wave system. If our unconscious ideas persist "engraved" in the cerebral proteins, these proteins are, like all molecules, only wave systems. The notion of "engraving" only has meaning for phenomena on our scale. Is it necessary then to admit that these wave systems are in turn "engraved" in a sub-matter? Physics has abandoned the old theory of the ether. The theory of the persistence of ideas as "engraved" in matter is no more justified than the theory of ether, considered as an indispensable material support for light waves.

THE DOUBLE COHERENCE OF MACHINES AND ORGANISMS

The autonomy of semantemes appears clearly in what we can call the double coherence of machines. (1) A machine is materially solid if it is well constructed. However, after being damaged, it is destined to rust and finally to go the cemetery for old scrap metal (unless by some chance it goes to the

Musée des Arts et Métiers). (2) But a machine, as a technical type, also has another coherence. It appears in the history of culture; it derives from another machine, it is reproduced, and perfected. We sometimes say of a technical *conception* that it is "solid." The history of machines is not the history of millions of assemblages of pieces of wood or metal, it is the history of technical ideas. A being of the (macroscopic) visible world only has a single sort of coherence: the degradation of a mountainous system, of a river or a rock, is irreversible. A used machine, however, can be repaired or replaced by another of the same type.

The "second coherence" of the machine is of course borrowed from the human organism: it is the human being who invents, maintains, repairs, reproduces, and progressively perfects it according to his ideas.

In general, however, every organism possesses a "double coherence" *by itself*. Each individual, as fully "solid" as he is, is destined to die: there are cemeteries for humans, just as there are for automobiles. But the animal or the man persists a lot longer as a species, by passing from individual support to individual support. The individual lends to the machine his own coherence, his "typical" or "ideal" immortality. Even if it is materially solid, without man the machine would obviously only be able to degrade, like a rock.

We clearly see the absurdity not only of understanding man through the machine but of resorting to the material solidity of his brain or of cerebral proteins in order to explain the solidity typical of man and his ideas—the source of the solidity "typical" of the machine. Moreover, embryogenesis directly demonstrates that the brain is remade for each individual.

All that remains for the materialist interpretation, then, is the resource of putting the continuity of organisms in the matter-form of genetic proteins, which in each embryogenesis would rebuild an individual organism, with its brain and preexisting cerebral memories, like micro-machines with programming and feedback. The *double* coherence of the organism would thus, strangely, be based on the *simple* coherence of its material reality. It is actually more rational to explain the double coherence of the living organism by the fact that it is a mixture of three modes of continuity: the material, the formal, and the semantic. The autonomy of the semantemes assures the typical continuity of man first and, indirectly, the machines that he constructs. The visible organism participates with some unobservable semantemes.

THE PROBLEM OF COPIES

The autonomy of semantemes with regard to space-time is the key to the indefinite persistence of living forms, which unlike the structures of the physical world, do not get lost—at least in their type—in the "background noise" that overwhelms everything. Contrary to appearances, it is also, first and foremost, what ensures the very multiplication of forms.

For Léon Brillouin, information (or in general, form) on the one hand degrades according to a generalized form of Carnot's principle, and on the other hand it can be multiplied by copying, "pedagogical" transmission, and so on.[27] A warm body loses its warmth, as it passes to a colder body (with loss of negentropy). Pierre, an informed man, passes his information to Paul, who, in receiving at least some of it, in general degrades it a little. The difference with the first case is that Pierre has lost nothing by communicating his knowledge. A professor teaches the Pythagorean table to his thirty students; each student remembers, or "gets," half of the table. The information has then been multiplied by sixteent (one for the professor, who always knows his table, and fifteen for the students).

There is a multiplication of information, although the principles of physics are satisfied by a degradation through the process. Moreover, according to Brillouin, the two phenomena—degradation and multiplication—are linked. There is a sort of compensation: the multiplication, the creation of order in the receiver, must be paid for by a consumption of negentropy. When a book is printed in ten thousand copies, each copy represents the same information as the manuscript, but negentropy was required to operate the printing presses, and each reading requires some expenditure of light or cerebral metabolism.

Bonsack sees things quite differently.[28] He maintains, against Brillouin, that the increase of information is only apparent. The cases of information of the teacher and of the students are not independent; they come from the same source and are linked together. Even if we add the information possessed by the teacher to the information acquired by the students, the total information does not exceed that of the teacher. In the case of a retransmitted message, according to physics the original always degrades. In the same way, every time a form is communicated, there is an echo effect, like a multiplication in mirrors. Between two parallel mirrors, an object appears multiplied to infinity: but is the information multiplied or increased? We can always carve out a "path of transmission" of a form or of some information, by as many segments as we like, and declare that each segment then multiplies the information. We can place a radio or television receiver in each cubic meter around the post of the emitter; I can theoretically place my eyes at any point in space and have a copy of the landscape. The (physical) images of an object are only ever the displaced and degraded form of this object. The division is arbitrary. From the point of view of physical reality, there is only a global phenomenon of extension-degradation. The cutting into slices, by receiving mirrors or by any kind of detector, multiplies nothing.

There is a true multiplication of information because the students, or the living and conscious readers, are not the same as mirrors or receiving apparatuses placed on the path of a flow of information in the process of degradation. In fact, the students can not only lose a lot less than 50 percent of the lesson of the teacher, but they can lose nothing at all. And if they are more intelligent than him, they can even improve and develop his ideas.

CONSCIOUSNESS AS INFORMATION SUPPORT

The sensations of living beings, as phenomena that are first of all physiological, can be degraded images. But when they become perception, that is, psychological information of which the support is the field of consciousness, meaning is reasserted in its autonomy. This is what the physiologists, after the psychologists, are discovering today. Naturally they are doing so in their own way, by opposing the eye or the ear, not to the active consciousness that understands their signals but to the brain.[29] Speaking of the brain as a material apparatus that performs the decoding of signals, entirely in keeping with their postulates, they in fact make it perform as if it were a mind (*esprit*) that makes hypotheses, trial and errors, reasons by implication, and hesitates if it does not have sufficient clues to draw a conclusion. It is not the first time in the history of the sciences that out of concern for positivity we fall into a magical conception worse than the one we wanted to avoid. Without attributing magical powers to the brain, let us instead directly examine the particular properties of consciousness as information support.

> A. Rather than "a consciousness," which risks being construed as a mythical personification, consider a field of consciousness. For example a visual field, when sensation, having ceased to be a physical image, in the physical space of waves and photons, has become a "subjective" image, but is not yet perception.
>
> According to Gregory, the eye supplies "coded information" to the brain: "We may take an analogy from written language: the letters and words on this page have certain meanings, to those who know the language. They affect the reader's brain appropriately, but they are not pictures. . . . No internal picture is involved."[30]
>
> This thesis mixes truth and falsehood. The true part is that the conscious visual field, whether obtained already at the level of the retina or only of the occipital zone, at an intermediary level, does not need to be looked at. It is, as we have often emphasized,[31] an "absolute surface," that is, a "surface" that is non-geometrical, a "surface-subject," which has no need of a point of perspective in a third dimension. However, it is false that the brain (or consciousness) somehow decodes

the information of perception and forms the *idea* of the object-sensation.[32] This would be to return to the conception of consciousness *facing* the object-sensation. The visual field of consciousness is an absolute and informed presence, before being secondarily interpreted according to all the themes and semantemes of the psychic organism.

Gregory sees the beginnings of a proof of his theory of coding in the recent discoveries of the physiologists David H. Hubel and Torsten Wiesel.[33] By presenting a cat with some "prepared" bars of light from various angles, they record in its brain some discharges of certain cells, when the bar is in a certain orientation, and of different cells when the orientation is different. Some different cells yet again are activated when there is a movement of stimulus in certain directions. Gregory concludes that there is no cerebral image but only some coded combinations of cellular activities. It is probable that there are some selective cellular analyzers of this type: color vision implies them, as does the sensation of movement in the periphery of the visual field without consciousness of a definite object. It is the same in hearing, where high and low notes are more finely analyzed than by the best physical receivers. However, we do not see why that would contradict the intuitively hard-to-reject notion of the overall field of sensation, with patterns and sequences, forms, and melodies, which are immediately expressive, if not meaningful or useful—the field from which certain elements are specifically selected, according to particular subordinate analyzers.

B. It's this character of the field of consciousness, of being a "subject-surface," which makes it an information support that has no analogy in the physical world.
1. The passage from a physical support to a conscious support for a form or piece of information produces some very particular effects. At the extreme, the conscious support can *give* a meaning or an expressiveness—that is, a semantics in the broad sense—to a physical form that doesn't have any. Consciousness transforms all form and even all appearance of form into information. It "reads" as a text even what is not a text. This fact has been of crucial importance in the history of humanity, and even in the history of life. Animals also "read"; they too create false meanings, but only according to their instincts. Man sees "meaning" everywhere, even in the physical world: in constellations, mountains and oceans, cataclysms and seasons, droughts and floods, as well as in the episodes of its life and those of other living beings.

This is not, by the way, purely an illusion: physical phenomena certainly have an inauspicious or favorable meaning *for him*. We understand that he is on the lookout for warning signs. The mistake is only in confusing the scowl of nature with the scowl of another human being.

The reading of physical phenomena can even become scientific, when the phenomena is channeled into derivations so that it can be recorded by instruments (such as the barometer, seismograph, or other recording device). Couffignal has proposed that we should consider such instruments as creators of information. But the information is only created when it is read. The decreases of barometers, by themselves, are part of the physical storm, in the same way that unperceived physical images of the sun are part of the sun. Measuring and recording devices are only tools of perception.

2. The field of consciousness as a "surface-subject" allows true conditioning or learning, not just their simulation. Contrary to the theses of Grey Walter and Dr. Sauvan, a spatial and temporal overlapping of two stimuli, the absolute and the conditional, is not enough to produce true conditioning. It is necessary that the animal "notice" this overlapping, and that it thus itself creates a true pattern, made of meaning, where the configuration is indissolubly tied to the thematic motivation. It needs a true conceptualization, as well as a true valuation. Grey Walter has since recognized, by the way, that conditioning and learning require an active, attentive, and selective patterning (which he naturally attributes to the brain, in keeping with his postulate): "It would be quite easy to make a computer-regulated auto-pilot for a car that would avoid obstacles and maintain an optimum speed, even in dense traffic, but to provide it with a visual recognition system that would distinguish red traffic lights from all other red lights, and would class them with the outstretched arm of a policeman, a H A L T sign, and other conditional signals—this would be a problem of a different order altogether."[34]

3. The surface-subject that is the field of consciousness allows invention, because it is a sort of canvas for "completion matrices," as in intelligence tests. If I see a set or series of forms, I can immediately notice their type or category. I can eliminate a form that stands out, put displaced forms in order, complete an incomplete series, fill in a blank, or invent the missing term from the perceived relationship. The printed material of a psychological matrix test clearly does not have any more tendency to be completed than a blank piece of paper has to cover itself with writing. Once transferred by perception to the field of consciousness as a universal canvas, the matrix has at least a chance of completing itself on its own.

Of course, consciousness is often stupid. Through years for the individual, and millions of years for the species, intelligent creation at the level of conscious perception (which is secondary in relation to the primary intelligent creation of the organisms of themselves by themselves) has remained difficult, and the word "immediate" hardly appears justified. It is nevertheless legitimate insofar as the intelligence of living beings comes directly from their consciousness. All animals, even inferior ones, spontaneously generalize and conceptualize. Machines behave "intelligently" solely through their assembly [*montage*], without being conscious, whereas the intelligence of living beings is inherent to their consciousness: their intelligent "set-ups" [*montages*] (for a detour, for example) come from their conscious "point of view."

STEINBUCH'S MATRICES

This difference appears very clearly with the interesting attempts of Karl Steinbuch and Uwe A.W. Piske to use so-called learning matrices.[35] These matrices are designed for the mechanical perception of forms and categorization. But Deweze has demonstrated that they could be considered as models of the mechanism of invention.[36] Steinbuch's matrices are true matrices in the mathematical sense of the word, that is, they are (physically realized) tables with connected rows and columns. The columns correspond to the characteristic criteria of the *structure* of objects of perception, while the rows correspond to the *categories* (or "meanings") to which the object belongs. During the learning phase, we apply a signal to the columns characteristic of the structure, and at the same time to the corresponding rows of "meaning," which creates a conditional connection. Then, the matrix can move to the active or "knowledgeable" phase: if we present it with a form (a letter, or the outline of a tree, house, or device), it identifies the category (or meaning). Conversely, if a category is indicated, it identifies some structural characteristics.

We can set up these matrices to operate with nonbinary but analogous signals, in a way that renders them indifferent to a certain number of variations. They then "recognize" a form, even if it is displaced, turned around, or distorted. To turn them into mechanisms of invention, it suffices to combine several matrices corresponding to some different objects or concepts, and to make them search systematically for the compatibilities or incompatibilities of these objects, either with a predetermined goal, or without predetermination—just as in invention, where it is important to not have a preconceived idea, which always risks being a blockage to a possible combination.

We see clearly—on this type of question, the intuition is indisputable—that human or animal intelligence does not function according to Steinbuch's matrices, that is, by a sort of scanning combining a double input. At least, intelligence in the elementary form of conscious intuition, or "insight" intelligence in Köhler's sense. When an animal makes a simple detour, or recognizes its nest; when a baby recognizes its mother in different clothes; when a man recognizes the nuts and screws in his toolbox, he uses the "absolute surface" character of the visual field. Form and signification are not dissociated, and they do not disappear into each other in the behavioral response. On the contrary, it is perfectly true that insight intelligence is quickly overwhelmed. Consciousness cannot always move directly from disorder to order. The scientist, and even the man in his ordinary life, then proceeds to examine possible combinations, first blindly then systematically, like a non-spatial labyrinth. He then secondarily sets up something like "Steinbuch matrices."

This is true for puzzle or "sliding blocks" problems, for example. That is, for all those tests with a simply matricial character, where it is difficult to visualize the intermediary combinations that constitute necessary detours toward the solution—thus the Passalong Test* or *Huit-test*, in which eight numbers, placed on a surface with nine places, one place being blank, must be slid in a way to change their order. The subject tested by this type of puzzle, after several practices, starts to discern and to form an idea of what is good or bad as an intermediary position. He senses the positions that are on the right track. It seems that a machine, in the same way, would be able to extract the strategic principles for using the intermediary positions, and do so more quickly and decisively than a human being.[37]

CORRECTION AND INVENTION

A related operation, almost indiscernible from invention, is correction, the restoration of degraded information according to its meaning. The filling-in of the blank of a test-matrix can be considered a correction, rather than an invention. Correction is what allows the true multiplication of information by

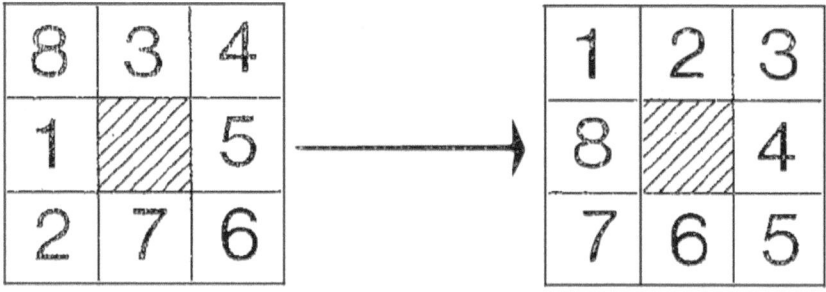

Figure 10.2.

restitutio ad integrum[38] of the spontaneously degraded text. The official thesis of cybernetics is that correction is not a true invention, because it necessarily uses a redundancy of the "message," which in general gives the same information twice or several times: on the one hand in a direct form, and on the other in a syntactic form. A language obeys rules (of spelling, syntax), which are types of constraint that represent prior information for the receiver. If I read "les chevaux galope," the information "plural" is only given to me twice by the article and the noun, instead of being given three times, as it grammatically should be.[39] But I don't have any difficulty restoring the plural of the verb in the text. A machine, as well as a consciousness, would theoretically be capable of restoring the text if it has the grammatical rules in its program.

But this is the same dilemma as that of automatic translation. Either there are no contextual clues, in which case the man as well as the machine must hesitate, or there are, and then a machine can use them as well as a man. Thus, the increase of information is only apparent. When I fill in a blank on a matrix by drawing or placing a *good* figure or *correct* word there, I simply use syntax in the absence of vocabulary. The power of consciousness here seems "to be based on a misunderstanding."[40]

The seemingly miraculous "powers" of the brain would in the same way be based on the enormous "redundancy" of its innumerable neuronal connections. As well as conceptualization, this enormous redundancy would explain the redintegration of a complete memory from very fragmented data, or despite brain damage. And Deweze even admits, with Forster, that it would be impossible to retain with any permanence a message that doesn't contain any redundancy.[41]

We believe that the "misunderstanding" here is rather on the part of the mechanist cyberneticians. As with translation, they conflate two concepts: that of the support, and that of the support-as-carrier-of-semantics. We can materialize the rules of spelling and syntax, and a machine can restore the plural of the adjective or the verb following the plural of the noun. It can also complete a test-matrix apparently following the meaning, if it works on a Steinbuch matrix, following "forced learning," or if it has readymade categories. For example, by drawing on its categorized vocabulary, it can complete "The opposite of little is . . . "

But consider for example a printing mistake, such as we find in Bonsack's book, which we have often cited, on the first line of page 84. In this phrase, printed in italics is "Un système abandonné à lui-même ne peut perdre de la valeur"; it is clear to the reader that the word "que" is missing between "peut" and "perdre." The printer makes Bonsack say the opposite of what he thinks.[42] However, the printed phrase is perfectly correct in its syntax. The reader restores following the context, especially after the preceding phrase: "The cause has at least as much value as its effects," because the reader thinks in the temporal succession implied in "the cause and its effects," and in the expression "abandonné à lui-même"[43] (implying "in time"). In short, he does indeed use the context, but the context in its meaning. Would a machine be able to accomplish this type of use? And above all, is it very clear here? Is it truly the only authentic scientific attitude, to appeal, with the mechanists, to "the structural richness and neuronal interconnections of the brain"? Scientifically speaking, this type of explanation makes us think of Empedocles rather than Newton.

A thesis such as Spearman's would appear more scientific. This thesis explains correction, translation, and test completion through a process of deducing the relation or theme, that is, an "eduction" or a "noegenesis." The process starts with incomplete terms and progresses to deducing first the relation and then the missing correlate. To reject a prejudice, such as in this case the spatialist prejudice, is to do excellent science. Especially insofar as this conception of "noegenesis" is in good accord with the great biological fact of the morphogenesis of the organism and the brain itself from a single cell, as well as its interconnections. The mechanists take this morphogenesis for granted, or they believe their geneticist colleagues are capable of explaining it scientifically from nucleic acids alone, with a microcybernetics.

Since both seem equally mythological, we might as well reject the hypothesis that all reality is synonymous with "the observable" (or "that which emits photons") in space, and instead adopt the postulate that the "observable" is connected with a "participable" domain of themes and meanings. From this perspective, brain-consciousnesses are two-way converters between physical space and the superspace of themes, just as organisms in development (embryos) are the site of a one-way conversion of specific themes into their spatial realization. Cerebral perceptive consciousness emerges naturally from formative organic consciousness as a continuing development.

The thesis of Deweze-Forster mentioned above, which says "that it would be impossible to retain, with any permanence, a message that doesn't contain any redundancy," may provide a plausible explanation of psychological memory as a property of the material brain. However, it makes the explanation of the formation of the brain with its "rich interconnections" even more difficult. Where is the redundant biological memory directing the formation of this brain? It would need to have even more complex redundant interconnections. Deweze's thesis is only the mechanistic disguise of the thesis, more in line with the experimental psychology of memory, which says that "No memory is possible without a retrieval of the semantic context, meaning, or themes, and without *psychological* redundancy."

INFORMATION AND ORGANIZATION

Cybernetics sometimes seems to hesitate between the word "information" and the word "form." This is natural, since one of its fundamental theses is the closeness—if not identification—of information and negentropy. All specific form, in opposition to fortuitous disorder or to a domain of variability, represents negentropy, and every information-message is a specific form in transmission and, if we can put it this way, in mission. Everyone seems to agree on this point: Couffignal, Bonsack, Brillouin, and Costa de Beauregard. "Nothing obliges us to limit ourselves to some messages[;] . . . we can understand these notions (the variability of a system and the specificity of a certain form) in any system and any form whatever. . . . Thus, a mosaic or an embroidery can be described (by counting some cases) as a sort of message of which the alphabet would be the range of colours and the specific form."[44] Similarly, Brillouin claims, "If information represents negentropy, it is the same for organization. . . . We now begin to see the relation with life. A living system is organised."[45] Similarly again, Costa de Beauregard considers organization as information, in the Aristotelean sense of active information, and he asks that we accept "in a similarly realist sense" the process of free action: information → negentropy, just as we accept the process of observation: negentropy → information.[46]

One must nevertheless be aware of the danger of speaking of "any form whatever," or of "negentropy" in general. The word "organization" is preferable. Of course, the mosaic, if not the embroidery, could be produced by the play of physical forces (artists today, we know, utilize a lot of chance). Its form would not appear to be less specific. We have spent billions to have some photographs of the lunar surface, as if the meteorite craters there formed a precious tableau. But true specificity, true form, implies an organization. For a statue, as for a message, there must be a third element, something "transcendent in relation to the system," which intervenes to put it in a well-defined state. It is only then "that it has a true specificity in relation to all possible states."

A well-coordinated crystal is not "specific in this sense,"[47] any more than is a soap bubble, or a liquid-steam system at a certain pressure. It results from a physical equilibrium. We know the misuse to which the notion of feedback,* or retroaction, has been put by confusing it with simple physical equilibrium, which has resulted in the absorption of all physical phenomena of equilibrium into cybernetics. A star does not pulsate through controlled, self-regulated oscillations like a steam engine equipped with a Watt regulator; a cave does not truly regulate its temperature, like an air conditioner. True cybernetic or biological organization demands a true regulation—at the level of active command. It is only then that the resulting form of this command can be put on the same plane as an information-message: it results from it.

Once again, cybernetics is inseparable from finality. Active regulation, in a thermostat for example, operates according to a true "message," with a semantics. If the thermostat's control mechanism is set at 18 degrees, it is not necessarily 18 degrees, it *represents* this temperature and as a result it contributes to producing or "realizing" it, like an end that determines its means. Organization cannot be confused with pure and simple negentropy.

This means that without exception, all organizations, like all messages, are directly or indirectly the work of living beings. Mechanical cybernetic arrangements are material, actual. They operate mechanically, but they come from a living being. It's the living person who, through interposed control, acts like the indispensable "third element," "transcendent in relation to the system."

And yet the ambition of cyberneticians is to understand living and conscious beings according to cybernetic models. The vicious circle is clear. The living organism explains *in reality* the cybernetic apparatuses. Then the cybernetic apparatuses are presented *in science* as explaining the living organism.

If we insist on applying cybernetic schemas to living beings and even man, then we will inevitably return to the excessively broad sense of the notion of feedback or retroaction: the living being will be a self-regulated form—but like a soap bubble or pulsating star, not like a thermostat or regulating machine. In sum, an inauthentic cybernetics will be an indispensable source of authentic cybernetics. The living person gives a meaning, but he does not have a meaning; he organizes, but he is not truly an organism. He chooses an effective model for a machine by comparing plans, but he himself results, not from a choice, but from a mechanical and blind natural selection. He invents, but he has not himself been invented.

CYBERNETICS AND BIOLOGICAL MORPHOGENESIS

The mechanist interpretation is absurd, with its claim to reduce all regulations to mechanical regulations, when in fact they are extracted from "ideal" or "valuing" regulations. Cybernetics, understood as the finality of machines, is not limited to an inauthentic cybernetics of pure physical equilibria. There may be a larger cybernetics, encompassing the space of mechanisms and the trans-spatial domain of values and meanings, one controlling the other, which in turn "realizes" it in space, through feedback.

In order to understand the morphogenesis and persistence of organisms, it is indispensable to consider that every living being is double, situated between two worlds, participating in space and in trans-space, such that it can play the double role of "assembling" and of "assembly." There are physiological feedbacks that only operate mechanically or chemically, and there are machines regulated internally that are present in all the organisms superior to viruses and single-celled organisms, as Cannon, Wiener, Stanley Jones, and all the physiologists have demonstrated. But physiology is not the whole of biology. It results from secondary assemblies. What is the agent of these assemblies? What is the active principle of morphogenesis or organogenesis?

Although cybernetics has always had a good relationship with physiology (it was born, in fact, from a crossing of physiology and mechanics), attempts by mechanist cybernetics to address either the morphogenesis of the individual[48] or of the species seem so far to have led to a dead end.

THE MORPHOGENESIS OF THE INDIVIDUAL

An idea of Ferdinand Gonseth is very illuminating here.[49] He emphasizes that there are two senses of the word "*ébauche*":[50] a weak and a strong one. It bears a weak sense when it is a first stage toward the realization of a final form that is known in advance (or already exists somewhere) and is essentially a sketch of the reproduction of an existing model. It has a strong sense when the draftsman, while being aware of being haunted by a vague idea, has no external or internal model and does not know in advance what will come from its first realizations, which are steps toward something not yet defined.

Now, the embryonic formation of a living being has something paradoxical about it: it needs to be classified as an *ébauche* in the "weak sense" developed by Gonseth: the finished structure, its adult version, exists somewhere. As we say, the formation is an event of reproduction. And yet, upon close examination of the epigenetic nature of development, the embryonic stages rather resemble a series of *ébauches* in the "strong sense": evidently, the formation does not copy anything. Experimental embryology confirms this. There are necessarily improvised regulations with only thematic orientations (toward a designable organ, first without precision, as cephalic, or dorsal, or caudal); "determinations," but with indeterminate content.

As for the *ébauche* in the weak sense, development at first appears to be explainable by cybernetics and mechanist models of controlling and programming material. The recent discoveries of the role of nucleic acids and their "messages" encoded in base sequences, as well as the role of their "messengers," which are the ribonucleic acids that enter the cellular protoplasm to construct specific proteins, have turned the idea that the entire organism is built on a genetic "organigram" into a dogma. It is believed that all its adult forms are inscribed in a coded form in the information contained in the genes of the initial cell, and that mechanisms of the cybernetic type, with various subordinate mechanisms and positive and negative feedback loops, ensure the transition from genetic information to the adult form, that is, from the program to its realization.

It is striking to see that embryologists, on the contrary, seem uncomfortable, while of course welcoming the discoveries of geneticists. This is because their experiments keep imposing on them the idea of the *ébauche* in the strong sense, which is incompatible with the notion of a fully predetermined program. One of the latest attempts to reconcile the irreconcilable (without mentioning that of Dr. Sauvan, which remains at the level of generalities), and for conceiving "the cybernetics of development" (it's the title of chapter 2 of his book) is that of C. H. Waddington.[51] It essentially consists in the metaphor of the "epigenetic landscape." It can be represented as shown in figure 10.3.

The ball, representing the state of the primordium *ébauche*, is not guided by rails but by valleys, more or less open, with thresholds, branches, and plateaus, where it can hesitate on its path, and then be determined by the action of a gene that helps it pass a threshold. It can accidentally (or through the intervention of the embryologist) leave the bottom of a valley, but it returns there through "homeorhesis" (rather than "homeostasis," since it is in movement). What then is the role of the genetic program? This program is represented by the strings from which hang the genes G, pulling the surface from the beginning, determining its bumps and valleys, and which can also deform it along the way through delayed intervention. It seems to be more a matter

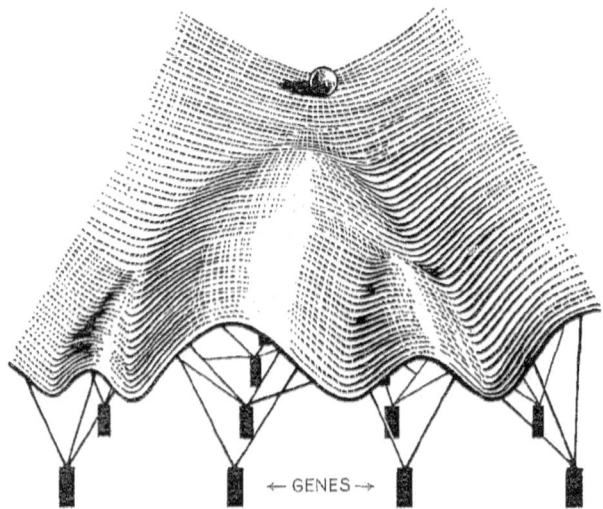

Figure 10.3. After C.H. Waddington (Figs. 4 and 5 combined)

of disguised preformation than of epigenesis. The programme is everything, and the ball has no initiative of its own.

Pure geneticists strongly advocate preformation and strict programming without the intervention of equilibrium of any kind. But can we then still speak of cybernetics? The reaction between the large molecules of enzymes are entirely different to ordinary chemical reactions, which find their equilibrium thanks to thermal agitation. Nucleic acids precisely control the sequencing of amino acids. "This is a mechanism as strict as a clockwork movement."[52] Jacques Monod has even spoken of a "neo-Cartesianism" of genetics. What he means by it is that the organism no longer appears today as a dynamic equilibrium, which would only have the type of stability of a steady-state* flame with a constant flow of materials, but as a sort of domino game with a strict spatial structuralism. Cybernetics would only appear in a way that is entirely residual, through genes or parts of genes that exist beside structural genes and are inaccurately called "regulatory." These act like repressors insofar as they block the activity of the operator genes.

It is difficult to think that this is the introduction of the cybernetic concept of feedback at the molecular level. The "repressor" or the "operon" is purely structural, like a cap on a bottle: it adapts itself to certain "sites" of the molecular structure, which it blocks or unblocks. However, a coaptation has never passed for a feedback equilibrium.

It is true that there remains the idea of code, which evokes cybernetics. But after a lot of enthusiasm, the cyberneticists have realized that the choice of this word has not been very adequate.[53] The correspondence (between the base triplets and the twenty amino acids) is imprecise. Above all we do not see where, and who, the decoder would be. Finally, the molecular structures do not *signify*; they act directly, like keys or plugs, or they act through their field.

Some English researchers (S. R. Pelc and M. G. E. Welton) have without doubt made the final blow against the metaphor of the "code," by discovering that the correspondence between the triplets and the amino acids is not arbitrary (like that of a sign to the signified). The three base triplets, assembled in space, fit together, or correspond spatially at a distance, with the acid that determines the triplet.[54]

But we are increasingly unable to see the connection between this domino game (where the dominos fit together according to their physical shape rather than by a player's "reading" of them) and the epigenetic developments of embryogenesis where on the contrary, the "causes" appear to be "signals." Waddington, in a recent book, expresses his concerns.[55] Molecular genetics can only claim to explain the macrostructures of the adult: "Some structures, essential for cellular development (revealed by the electron microscope) already do not appear like the direct product of the gene-enzyme systems, but rather as secondary elaborations." We can say the same thing, and even more so, about macrostructures. The chemical explanation (by the "inducers") explains nothing. What counts is the response of tissues, their "competence," as the embryologists say, reacting to the "signals" of the inducer. We would say that it is their own semantemes, simply awakened by the signals. Waddington protests against the exclusive enthusiasm for molecular genetics and calls for greater effort at the fringes of fundamental embryology.

To sum up, mechanist cybernetics fails in every way in biology (except physiology, naturally).

a. There is no mechanico-chemical feedback in development.
   Development is, paradoxically, a reproductive epigenesis. It is conceivable as a regulated invention; it requires the hypothesis of a control and a feedback *of some sort*, but a control transversal to space, with a thematic character. The chemical substances of "inducers" appear to be only the "supports" of a semantics.
b. On the other hand, it fails again when it considers development to be strictly programmed according to a genetic code. The American physicist Walter M. Elasser is consistent with the suggestions of Niels Bohr, Pascual Jordan, and Brillouin—and goes against Erwin Schrödinger's theses—in demonstrating that it would be arbitrary to postulate a passage from "information" stored in the genes to the structure of the adult, which would be determinist in the sense of classical physics.[56] The idea of a "Laplacian" observer—who would verify this new preformationism by "enquiring" about the way in which the information contained in the molecules of DNA, then of RNA, determine large scale organization through various intermediaries—remains in the domain of pure unverifiable theory. Just as much as the idea of "telegraphing" a man over a distance by examining his molecular structures and coding them into a message, which a receiver would decode in reconstituting the man, as Brillouin has demonstrated. As for the chess-playing automaton, the operation of calculators and electronic readers, even at fantastic speed, would take more than a hundred centuries in the best hypothesis.[57]
   The "stored" program of genetics is like a book that everyone swears is extremely interesting, but that no one has ever been able to read, and of which it could be proven that no one could ever read.

Note that it is not the structure of some chains or helices of DNA that is "unobservable," or the event of their duplication or their molding, it is the *determinist* dependence of the organic structure of the adult on these DNA structures, which are supposed to command it, or to contain it *in nunce*.

However, numerous indications show that observable large organisms, with their primary and secondary consciousness (meaning that these organisms are "absolute" surfaces or domains, possessing their own form, without needing to observe themselves by scanning from an exterior point) emerge in temporal continuity by genidentity, through the virus or other living molecules from microphysical realities. Microphysical domains are, like organisms, domains of delocalization, or internal detemporalization. Through their multiplicity they form classical space-time with secondary statistical laws. But in themselves they are below these laws—which explains why the large organisms themselves appear to be beyond these laws.

We understand very well, then, the failure of cybernetics when confronted with life and consciousness. The cybernetic models all borrow from classical, macroscopic physics. They are just as incapable of explaining the lowest living individual as the model of the "impact between billiard balls" is incapable of explaining the electron, the neutron, and all the elementary particles and their interactions.

Cybernetics succeeds in physiology, *at the level where, precisely, the living individual uses large statistical phenomena secondarily*, at the level of the organic machines that it assembles as internal auxiliaries. Here it is only natural that the cybernetic models, and in general, external machines, correspond. The machines, whether internal or external, put to work inorganic materials, or materials derived from the organic domain, channeling the flow of other materials or energy, and sometimes using the currents derived as guiding feedback.

We understand just as well how the notion of mechanical feedback, using negentropy to create informative order, loses its usual sense in the domain of microphysics. We can guide a D.C.A. cannon by a radar that observes the target and guides or corrects the aim, by stopping the cannon in the right position. This stopping requires a "microscopization" of the "disorder."[58] But Maxwell's demon, which is supposed to observe the molecules and choose the faster ones, by letting them pass through its trapdoor—in the same way that Szilard's demon, which uses a unique molecule in a cylinder with a piston without the friction that this molecule would generate—cannot stop these molecules, by making microscopic a disorder that is already microscopic. Without its author appearing to perceive it clearly, this excellent analysis condemns all microscopic cybernetics, and so, every mechanist cybernetic explanation of life. The Maxwellian or Szilardian demons (if we can speak of demons) that work in the elementary organisms to choose molecules and combine them, surely do not work according to the laws of classical physics.

THE MORPHOGENESIS OF SPECIES

Interestingly enough, although there was an excessive haste to incorporate cybernetic models into the morphogenesis of the individual, that urge was abstained from when it came to the morphogenesis of the species. According to neo-Darwinism and genetics, this morphogenesis is the work of unidirectional mutations, without any recurrent guidance from the results of these mutations. This is because there is no inheritance of acquired characteristics through which the needs of adaptation to the environment would be able to guide the mutations in turn. Living species do not adapt—they are transformed, by chance genetic mutations. In that, they are inferior to the *Machina docilis*, the Homeostat, or the canon's radar. Genetic mutations produce *arbitrary* adult structures, and species only adapt to the environment at the cost of the enormous eliminations of natural selection, without true regulation. Species are even inferior to *Machina labyrinthea*, since in the latter the *correct* path is recorded in memory and control mechanisms on the basis of successful attempts.

This thesis is perfectly implausible, and this implausibility is increasingly troubling to the biologists. We will mention only one well-established fact. It is often observed that a particular environment induces some phenotypic (that is, nonhereditary) modifications in a species that are adaptive, and which exactly resemble aberrant forms of the same species that are genetically induced and fixed. More generally, besides these hereditary characteristics of which we do not see how they would be able to result from an active and self-regulated adaptation (for example, the transparency of the cornea), there are others that exactly resemble the results of this type of adaptation (for example, the callosity of the underside of the ostrich, to the place where the animal squats).

In addition, after having rejected Lysenko and the Soviet school, a lot of Western biologists today are working to reintroduce the idea of a control feedback, coming from individual adaptation and modifying the genes, which thus would not work blindly and in a single direction. But currently the models proposed according to the mechanist postulates are quite unconvincing.[59] It seems once again that (1) feedbacks or couplings of some sort are indispensable, and (2) these feedbacks cannot be exclusively mechanical.

CONCLUSION

Apart from its general interest as a mobilizer of ideas, the partial successes and failures of cybernetics are paradoxical. The proverbial axiom "He who can do the most, can do the least" does not apply to it. Or, we must revise our notions of more and less, easy and difficult. Cybernetic machines or schemas easily imitate logical reasoning and calculation. They do not just imitate or simulate; they embody them in a way that is more perfect than the human brain is capable of. But it is much more difficult for them to imitate conditioning, association, and perception, which they only manage to simulate. And they do not even simulate sensation.

They can serve as auxiliaries for the difficult decisions of managers, through operational and strategic calculations, such as for the difficult maneuvers of a spaceship. But they cannot choose between two colors of furniture.

Translation machines are still in their infancy, or even in limbo. A machine can successfully translate, from Russian into English, some sentences that are already mechanical, such as "*Worker fraternal socialist country express unanimous support position Soviet government expounded Kroutchev in Paris.**" But for a Russian poem, it gives this:

> *No, not in reception room, no, not house*
> *Day, And Two. And five.*
> *Where it, predrayispokoma*
> *It in way again.**

How can we understand that these are the complaints of a fiancé whose lover is too involved in politics to be at home very much?[60]

Paradoxically, however, if the machine cannot translate poems, it can make them easily, and ones which are difficult to distinguish from those of authors who are much admired. Albert Ducrocq's "Calliope" (with, it's true, a vocabulary that is quite strictly selected) produces some beautiful passages:

> *Un rideau de plantes rouge meuble l'éternité . . .*
> *Le hérisson avance péniblement, le corail rêve . . .*
> *Tous aiment peindre la terre.*[61]

Machines for composing music, for painting, for sculpture, and especially for inventing some new forms of art—where time intervenes in a way more purely aesthetic than in cinema, such as in the "luminodynamism" or "chronodynamism" of Nicolas Schöffer, which uses rotors, punched plates, light-bending mirrors, and so forth—has increasingly gained the attention of the public, and especially of artists. Clearly, we can remark (1) that the kaleidoscope has been known for a long time; (2) that the poems or spectacles, which seem to be produced by machine, are in reality made by the consciousness of the reader or spectator, like magical houses and islands in the clouds; and (3) that such a machine could very well have been invented by a Pascal in the seventeenth century (Raymond Lull had already invented a machine for philosophizing in the thirteenth century), who would have been willing to call the product of this machine a "poem" or a "painting," but whom the public would have ignored. Today, the public is prepared—by poets and painters, before there was any cybernetics—to call "poems" sequences of words whose evocative power is indeterminate, and "paintings" any marks on a canvas.

It would not be difficult, by employing not even the most modern machines, but Lull's *Ars magna*, or the machine of the professor of Laputa (from *Gulliver's Travels*), to make some small cubes, which could be turned by a crank handle. On the faces of these cubes would be printed expressions such as "valid option"; "begin the dialogue"; "confrontations"; "obligatory moves"; "lead to . . . "; "reconsider"; "open to the future"; "settle"; "structural reforms"; or "be destined to . . . " With this, it would not be difficult to produce a speech that appears to be very good. But, of course, only for a public open to the future, which would be willing to hear such a message emerging through such options.

Another significant achievement of cybernetics is that information machines, as well as power machines, surpass the powers of man. Deweze gives a striking example, drawn from the "perceptron."[62] In the course of correcting learning, the human instructor is alleged to make mistakes in their corrections 30 percent of the time; the performances of the "perceptron" are ultimately better than those of the instructor. Louis Couffignal shows that logical machines can carry out operations more complicated than those ever performed by humans. By combining some information supports, the machine can arrive at a semantics that is valid if the information is unambiguous, and that a person would be incapable of interpreting. Lewis Carroll's cascades of syllogisms, of which the machine easily draws the conclusion, is already disturbing to the mind. The means the Homeostat uses to arrive at its equilibrium can be very difficult to identify and to predict. A machine that is designed to achieve an end given by a human being can resort to some means that are unexpected, while still being consistent with its assembly. This could eventually be catastrophic, as in the story "The Monkey's Paw,"[63] where the wish to have two hundred pounds for a mortgage payment is very quickly realized—but in the form of a pension, provided for the accidental death of the son of the house. A boiler, set to provide me with a certain temperature, can similarly obey my wish to be warm—by asphyxiating me.

But let's be clear. This superiority of the machines does not mean that the machine can do without human beings, or more precisely, that it can do without a framing consciousness of some sort. This point is absolutely crucial. To misunderstand it would lead to the most serious errors, both biological and psychological. Aurel David enumerates what he considers to be a series of erroneous assertions by cybernetics. Along with "A machine will never know more than its constructor knows" and "There will never be a machine capable of such and such action," he adds another assertion that he considers to be equally false: "A human being will always be needed to construct, guide, monitor, and repair the machine." He reminds us (1) that the interventions of a repairer are comfortably classified with the actions that are more easily mechanizable; (2) that human surveillance is advantageously replaced by automatic surveillance—this is the essence itself of cybernetics; (3) that there are some automatic textile factories that work night and day in complete darkness (we only light the room of the machines for visitors); and (4) that a machine can construct a machine, by inventing even the "means of construction," provided that a goal is given to it. For example, a machine (set up for self-learning) that manufactures screws, which initially had no mechanism intended for this production, and to which we only give a finality by presenting a screw as a model and forcing the machine (through its primary assembly) to "punish itself" when its processes have been ineffective.[64]

Even if we don't condemn the notion of "self-learning," about which Deweze is less optimistic than David, it is evident that a human being (or consciousness) is always indirectly active and framing, as the "giver of finality." We can place regulating or repairing machines on other machines, but we do not change their mode of existence, which is material like minerals, or formal like physical equilibria. Only a consciousness can give them their "second mode of existence," as technical idea.

A machine is understood through involvement, according to the goal of its constructor, and not by a pure description of its parts as physical realities. If we claimed to describe a certain assemblage of metal, mica, and porcelain in the motor of an automobile (the spark plug), without knowing that it acts as a device designed to produce sparks,[65] this absurd positivism would amount to nothing.

Aurel David's thesis is not serious. The important thing is not that a human being is present, in flesh and blood, beside the repairing machine. If he does not understand its mechanics, his presence will be useless. The important thing is that he exists *as living support of the technical idea*, according to which the machine was constructed, and that he is ready to intervene and repair it according to this idea. The "Surveyor"* on the moon would be 400,000 kilometers from man. But it would be dependent on man nevertheless.[66]

Machines, with or without an added system of regulation and repair, do not have their own consistency. It is characteristic that the internal machines themselves, the organs, only subsist through the organic primary consciousness that frames them. A superior animal is mortal because it can no longer repair the too-extensive injuries to its organs. A heart is not immune to heart diseases, even though it has additional regulatory mechanisms on top of its own beating. By contrast, a unicellular organism is virtually immortal without any machines because its "body" is only the observable part of its domain of absolute surface. And the large organisms would have to again pass through the unicellular phase, by abandoning all their machinery, in order to reproduce and keep their specific genidentity.

If an epidemic caused humanity to disappear, with the exception of several people on a desert island, they would repopulate the earth thousands of years later. The automatic textile factories would have long since turned to dust, while the new humanity would rebuild new technical lineages, like a single germinal cell rebuilds the machinery of an adult body.

The controversy between mechanist (or dogmatic) cyberneticians and critical cyberneticians has taken some stereotypical forms. We can schematize it with a dialogue between D (the dogmatist) and C (the critic).

D—You state the obvious by recalling that machines and mechanisms are neither conscious nor living. We know that just as well as you. Everyone knows it. Only science fiction authors have seemed, for the amusement of their readers, not to know it. Perhaps we should add a few eccentric philosophers, whose intentions aren't entirely clear, and perhaps they just want to amuse themselves.

C—Very well, but then why present your mechanical "models" as biological and psychological discoveries? And above all, why accuse the skeptics and critics of not having the scientific spirit, and of clinging to myth and superstition?

D—We don't present these models as discoveries, but as work instruments. We've even proposed to always add a letter of discrimination, to distinguish memory M from memory H, learning M from learning H. We're always ready to abandon our models as soon as facts disqualify them.

C—You know the saying attributed to a wife, living only with her old husband: "If one of us dies, I will go and live in Paris." You do not see any other solution, or any other survival, than your own. You're "always ready" to abandon mechanist models. But as soon as we suggest that this moment seems to have come, you decry the lack of scientific spirit.

D—It is indeed unreasonable to be discouraged. Let's perfect our model to the point of exhaustion and wait until we have discerned a truly irreducible residue; only then will we think about reintroducing this residue into science.

C—If that's your method, it's bad. This "residue" could be a fundamental element, that would require a complete overhaul of our conceptions. In fact, the physics you use is already outdated. You act as if microphysics doesn't exist. You treat organisms as machines made from homogenous material, and when you descend to the microphysical scale, you introduce some notions borrowed from large-scale phenomena.

D—Our models of learning, conditioning, and conceptualization work. That's all that matters.

C—They're simulators, not models. They don't resemble the biological and psychological facts. They only resemble the false idea that you have of the biological and psychological facts. You apply an outdated physics to a biology and psychology that many biologists and nearly all psychologists no longer want. Too often cybernetics has been only a pretext for the resurrection of the more indefensible forms of scientific dogmatism and even totalitarianism. Disguised like this, mechanism has regained a good conscience. Acting like a new Descartes, Albert Ducrocq today explains everything under the name of cybernetics: life, Man, the universe (and even a little more, since he claims to explain the *necessary* origin of the universe). He (or his editor) presents "the sum of human knowledge that has allowed the discovery of that without which the earth and life could not have emerged."[67]

D—The comparison is flattering for Ducrocq.

C—The truth is that I don't have such a high opinion of Descartes, as author of the World, as physicist "doing everything," whose place would have been between Empedocles and Democritus rather than between Galileo and Newton.

This totalitarianism has never succeeded: it consists in the forceful application in all areas of science what has only been partially successful in one. Besides, your psychology is not only bad but also nonexistent. You have only contributed to equipping psychologists—who absolutely want to be treated like scientists—with a vocabulary that is pretentiously mathematical and falsely precise. Thankfully, linguistics is solidly established and has its own autonomy. Otherwise, you would have reduced it solely to something like Zipf's law, and to research on "the temperature of vocabulary."

D—This is better than seeing linguistics left in the hands of phenomenologists and existentialists, biology left to vitalists, and psychology to neo-animists who put "a ghost* in the machine." But seriously, what gives me confidence is that our models, which are obviously very far from perfect, continuously improve themselves. They grasp reality more and more accurately; they conform to it asymptotically. In the process, they force the various specialists to analyze the facts they are dealing with more carefully.

C—This asymptotism is as much my principle as yours. If, as I believe, and as everything suggests, organisms and their primary and perceptive consciousness emerge directly from the microphysical domains that have managed to frame and utilize crowd phenomena by setting up one relay on another, there is no doubt that in the balance of forces and materials, their action is almost imperceptible. It is like the action of a pilot, or rather the pilot's nerve cells, in the maneuvers of an ocean liner or the Apollo space rocket.

Your descriptions, even though they may be quantitatively asymptotic, are nevertheless false and misleading.

When we abandon the mechanist postulates, we do not find ourselves in the presence of cybernetics *minus* something. On the contrary, we have cybernetics *plus* a powerful method for exploring the problems of life and consciousness.

D—But considering any biological and psychological performance, you could never demonstrate to me that a machine could not do it.

C—But you can never demonstrate to me that a machine will be able to do it. We are even.

Since we have agreed that a robot will never be alive and conscious, we are in agreement that we will never explain life and consciousness with a mechanistic cybernetics.

Notes

TRANSLATOR'S INTRODUCTION

1. Andrew Iliadis, "Mechanology: Machine Typologies and the Birth of Philosophy of Technology in France (1932–1958)," *Systema*, 3.1 (2015): 137; and Iliadis, "Introduction to Ontologies of Difference: The Philosophies of Gilbert Simondon and Raymond Ruyer," *Deleuze Studies* 11, no. 4 (2017).

2. I will refer to this as "mechanical information," or the "mechanical model" of information. This is not a commonly used expression, by either the cyberneticians or Ruyer, and is somewhat reductive, since Shannon and Wiener's information theory aims to capture more than what might usually be deemed "mechanical." It does however capture well a distinction Ruyer wants to make, in a short-hand way, so I will use it here as an expedient. "Mechanical" should be understood both in the sense of applying to machines and engineering, and of statistical mechanics.

3. Luciano Floridi, *The Philosophy of Information* (Oxford: Oxford University Press, 2011).

4. Guowu Li, "Information Philosophy in China: Professor Wu Kun's 30 Years of Academic Thinking in Information Philosophy," *TripleC* 9, no. 2 (2011): 316–21.

5. See for example Ruyer's following articles: "Les informations de présence," *Revue Philosophique de la France et de l'Étranger* 152 (1962): 197–218; "La quasi-information. Réflexions en marge de deux ouvrages récents," *Revue philosophique de la France et de l'Étranger* 155 (1965): 285–302; "Quasi-information, psychologisme et culturalisme," *Revue de métaphysique et de morale* 70, no. 4 (1965): 385–418; "Cybernétique et informatique' in *La philosophie contemporaine, vol. II: Philosophie des sciences*, ed. R. Klibansky (Florence: La Nuova Italia, 1966). This is a non-exhaustive list but indicates some of the main articles in which Ruyer's reflections on cybernetics and information were continued.

6. See especially the introductory section of Raymond Ruyer, Tano S. Posteraro, and Jon Roffe, "Instinct, Consciousness, Life: Ruyer contra Bergson," *Angelaki* 24, no. 5 (2019): 124–47; Daniel W. Smith, "Raymond Ruyer and the Metaphysics of Absolute Forms," *Parrhesia* 27 (2017): 116–28; and Mark B. N. Hansen's "Introduction: Form and Phenomenon in Raymond Ruyer's Philosophy," in Raymond Ruyer, *Neofinalism*, trans. Alyosha Edlebi (Minneapolis: University of Minnesota Press, 2016).

7. On the history of this reception, see Céline Lafontaine, "The Cybernetic Matrix of 'French Theory,'" *Theory, Culture & Society* 24, no. 5 (2007): 27–46; and Bernard Dionysius Geoghegan, *Code: From Information Theory to French Theory* (Durham and London: Duke University Press, 2023).

8. Jean-Hugues Barthélémy, *Penser l'individuation: Simondon et la philosophie de la nature* (Paris: L'Harmattan, 2005), 21. For a (non-exhaustive) list of some of the brief references to Ruyer's book by other French intellectuals (including Lacan, Merleau-Ponty, Deleuze, and Piaget), see Alix Veilhan, "Raymond Ruyer et la cybernétique," *Philosophie* 149 (2021/2): 55, note 91.

9. Raymond Ruyer, Tano S. Posteraro, and Jon Roffe. "Instinct, Consciousness, Life: Ruyer contra Bergson," *Angelaki* 24, no. 5 (2019): 126.

10. Norbert Wiener, *Cybernetics, or Control and Communication in the Animal and the Machine*. 2nd Ed. (1964) (Cambridge: MIT Press, 1991).

11. Quoted in Philippe Gagnon, *La réalité du champ axiologique: Cybernétique et pensée de l'information chez Raymond Ruyer* (Louvain-la-Neuve: Les Éditions Chromatika, 2018), 569.

12. This volume, 2–3.
13. This volume, 8.
14. This volume, 10–11.
15. This volume, 29.
16. This volume, 82. Wiener, *Cybernetics*, 91; 62–63.
17. In the section of the introduction titled "Perpetual Motion of the Third Kind."
18. This volume, xxiv.
19. This volume, 6–7.
20. This volume, 7.
21. This volume, 6. Proposed by Sadi Carnot in 1824, Carnot's Principle is a theorem that specifies the upper limit of efficiency of a heat engine with mathematical precision.
22. This volume, 7.
23. This volume, 5.
24. This volume, 71.
25. This is of course a point that contemporary developments in information technologies, such as natural language processing (NLP), might cause us to question. See the last part of this introduction for some general considerations concerning the "datedness" of some of Ruyer's arguments.
26. For a development of these various ideas, see Raymond Ruyer, *Neofinalism*, trans. Alyosha Edlebi (Minneapolis: University of Minnesota Press, 2016).
27. While nevertheless strictly qualifying this "Platonism." In *Neofinalism* he insists, "The region of essences and themes should not be situated in some mythical geography, like the one that amused Plato. It can be reached through positive descriptions of a certain number of psychological facts, all of which reveal the same structure" (see page 124).
28. This volume, 74.

29. Raymond Ruyer, "Le problème de l'information et la cybernétique," *Journal de psychologie normale et pathologique* 45 (1952): 402. All translations from cited French texts other than the present book are wholly my own.

30. Ruyer, "Le problème de l'information et la cybernétique,' 403.

31. Ibid., 402–3.

32. Ibid., 404.

33. Ibid., 404–5. My account here has followed Ruyer's brief presentation in this paper. Similarly brief accounts are given at various places in *Cybernetics and the Origin of Information*. A much fuller development of these points about biological reproduction is given in Ruyer's major treatment of the topic, *The Genesis of Living Forms*, trans. Jon Roffe and Nicholas B. de Weydenthal (London and New York: Rowman & Littlefield International, 2019).

34. Ibid., 406.

35. Ibid., 406–7.

36. Raymond Ruyer, *Paradoxes de la conscience et limites de l'automatisme* (Paris: Albin Michel, 1966), 235.

37. Ruyer, "Le problème de l'information et la cybernétique," 418.

38. A few selected examples are N. Katherine Hayles, *How We Became Posthuman: Virtual Bodies in Cybernetics, Literature, and Informatics* (Chicago: University of Chicago Press, 1999); Jean-Pierre Dupuy, *On the Origins of Cognitive Science: The Mechanization of the Mind*, trans. M. B. DeBevoise (Cambridge, Massachusetts: MIT Press, 2009); Andrew Pickering, *The Cybernetic Brain: Sketches of Another Future* (Chicago: University of Chicago Press, 2011); and Geoghegan, *Code*.

39. See for example the dialogue that closes chapter 10.

40. Luciano Floridi, *The Philosophy of Information* (Oxford: Oxford University Press, 2011), chapter 2.

41. Floridi, *Philosophy of Information*, 35.

42. Mark B. N. Hansen, *New Philosophy for New Media* (New York: MIT Press, 2004), 79.

43. Raymond Ruyer, "Quasi-information, psychologisme et culturalisme," *Revue de métaphysique et de morale* 70, no. 4 (1965): 386.

44. See Floridi, "Semantic Conceptions of Information," *Stanford Encyclopedia of Philosophy* https://plato.stanford.edu/entries/information-semantic/ (revised January 7, 2015).

45. In chapter 10, "The Problems of Cybernetics in 1967," which is a major addition to the second edition of *Cybernetics and the Origin of Information*, Ruyer notes that the word "cybernetics" had become less used over the last decade, while research in its central topics, the technics and theory of informtion, had continued to expand. This volume, 125.

46. See John Searle, "Minds, Brains and Programs," *Behavioral and Brain Sciences* 3 (1980): 417–57; and Ned Block, "Troubles with Functionalism," *Minnesota Studies in the Philosophy of Science* 9 (1978): 261–325.

47. See Ruyer, "Le problème de l'information et la cybernétique," 393–97.

48. Floridi, *Philosophy of Information*, 43.

49. Ibid., 43.

50. Floridi foregrounds the provisional and non-exhaustive nature of his list, and states that he "would not mind reconsidering which problem belongs to which area or further problems that need to be addressed." *Philosophy of Information*, 30.

51. Georges Chapouthier, "Information, structure et forme dans la pensée de Raymond Ruyer," *Revue Philosophique de la France et de l'Étranger*, 203, no. 1 (2013): 21–28.

52. This was indeed broadly the case with Simondon, who, Barthélémy notes, found Ruyer's metaphysics too "vitalist" and "spiritual." *Penser l'individuation*, 128.

INTRODUCTION

1. TN: "control by" in second edition only.

2. Norbert Wiener, *Cybernetics, or Control and Communication in the Animal and the Machine* (Cambridge: MIT Press, 1948), 49.

3. TN: Samuel Butler, *Erewhon: or, Over the Range* (London: Penguin Classics, 1985), 203.

4. TN: Here and following, "*pattern*" in the first edition is replaced by "support" in the second edition.

5. See chapter 5.

6. Here and following, "*pattern*" is replaced by "structure" in the second edition.

7. TN: *effets de clé*—Ruyer is suggesting a turning and unlocking effect.

8. TN: See Warren S. McCulloch and Walter H. Pitts, "A Logical Calculus of the Ideas Immanent in Nervous Activity," *Bulletin of Mathematical Biophysics*, Vol. 5 (1943), 115–33, reprinted in Warren S. McCulloch, *Embodiments of Mind* (Cambridge: MIT, 1965), 11–18.

9. "Structures" in second edition.

10. TN: Ruyer is here drawing a correspondence between the first two types of machines that Wiener identifies and the first two types of perpetual motion as standardly classified according to the thermodynamic principles that they violate. "Perpetual motion of the first kind" refers to a machine that would work indefinitely without any energy source, which violates the first law of thermodynamics (i.e., energy cannot be created or destroyed). "Perpetual motion of the second kind" refers to a machine that would work indefinitely with a single (thermal) energy source, which violates the second law of thermodynamics (i.e., the increase of entropy). There is no consensus on the meaning of the term "perpetual motion of the third kind," but it is often used to refer to machines that would purport to work indefinitely because all friction is eliminated. Ruyer's suggestion that information machines can be more efficient than motor machines, and are not restricted by Carnot's law, heads in this direction; but his main point is beyond this, since information machines would need not simply to reproduce given information indefinitely, but create new information.

11. Here and following, "structure" in the second edition.

12. Edmund C. Berkeley, *Giant Brains; or, Machines That Think* (New York: John Wiley and Sons, 1949), 180.

13. TN: Henri Bergson, *The Two Sources of Morality and Religion*, 268, translation modified (the standard English translation reads, "We must add that the body, now larger, calls for a bigger soul, and that mechanism should mean mysticism." Ruyer, apparently citing by heart ("ce corps massif attend un supplément d'ame"), slightly misquotes the original French text, which reads, in full, "Ajoutons que le corps agrandi attend un supplément dâme, et que la mécanique exigerait une mystique." Henri Bergson, *Les Deux Sources de la morale et de la religion* (Paris: PUF, 1984 [1932]), 335.

14. Georges Friedmann, *Problèmes humains du machinisme industriel* (Paris: Gallimard, 1946), 173.

15. TN: Here and throughout the book, the first edition has "*feed back*" (italicized, in English, with a break), while this term is frequently replaced in the second edition by "rétroaction." We will use the English term standard in the cybernetic literature, "feedback."

16. Here and in the rest of this section, the second edition, as well as the first, has "*feed back*."

17. Henri Dubreuil, *Standards: Le travail américain vu par un ouvrier français* (Paris: Grasset, 1929).

18. Henri Frankfort, *Before Philosophy: The Intellectual Adventure of Ancient Man* (New York: Penguin, 1946).

19. Olaf Stapledon, *Last and First Men*, 5th ed. (London: Metheun, 1934), 218 (chapter X).

CHAPTER 1

1. TN: This last sentence is added in the second edition only.

2. The term "of logical reasoning" in second edition only.

3. George Boole, *An Investigation of the Laws of Thought* (New York: Dover, 1958).

4. See Edmund C. Berkeley, *Giant Brains* (New York: John Wiley and Sons, 1949), 152; and Louis Couffignal, *Les machines à penser* (Paris: Les Éditions de Minuit, 1952), chapter VII.

5. TN: Claude E. Shannon, "A Symbolic Analysis of Relay and Switching Circuits," *Transactions of the American Institute Electrical Engineers* 57, no. 12 (December 1938): 713–23.

6. Italicized in second edition only.

7. TN: A portmanteau of Voice Operating Demonstrator, the *Voder* was the first electronic human speech synthesizer, invented by Homer Dudley for Bell Telephone Laboratories in 1937–1938.

8. Berkeley, Giant Brains, 22.

9. Clark L. Hull, "An Automatic Correlation Calculating Machine," *Journal of the American Statistical Association* 20 (1925), 522–31.

10. Charles Spearman, *The Nature of Intelligence and the Principles of Cognition* (London: MacMillian, 1923), chapter VII.

11. Pierre David, *Le Radar* (Paris: Presses Universitaires de France, 1949), 98.

12. Ibid.

13. On self-regulation and feedback, see in particular Pierre de Latil, *La Pensée artificielle* (Paris: Gallimard), 1953. Translated as *Thinking by Machine: A Study of Cybernetics* (Boston: Houghton Mifflin, 1957).
14. In the second edition, it is "The feedback is negative."
15. The phrase "which races or returns to zero (runaway*)" is in the second edition only.
16. Herman J. Jordan, "La conception naturaliste du monde dans ses rapports avec la méthode dialectique ou synthétique en biologie," *Recherches philosophiques* 1 (1932), 191.
17. Walter B. Cannon, *The Wisdom of the Body* (New York: W.W. Nortin & Co., 1939). Ruyer cites the French translation: *La sagesse du corps* (Paris: Édition de la nouvelle revue critique, 1946).
18. TN: A type of compass.
19. Clark L. Hull, *Principles of Behavior* (New York: Appelton-Century, 1943), 27.
20. W. Grey Walter, *The Living Brain* (London: Gerald Duckworth & Co., 1953), chapter 5.
21. Norbert Wiener, *Cybernetics*, chapter 6, and W. Russell Brain, "The Concept of the Schema in Neurology and Psychiatry," in *Perspectives in Neuropsychiatry*, ed. Derek Richter (London: H. K. Lewis & Co, 1950). Ruyer cites the French translation of the latter: W. Russel Brain, "La notion de scheme en neurologie et en psychiatrie," in *Perspectives cybernétique en psychophysiology*, trans. J. Cabaret (Paris: Presses Universitaires de France, 1951), 33.
22. This English term appears, here and in the following passages, in the first edition only. The second edition has "*balayage*," with a footnote indicating "or *scanning*" on the first mention.
23. "Group scanning" in English is in the first edition only.
24. Brain, "La notion de scheme," 39.
25. Ibid., 43–44.
26. The first and second editions give different references here. In the first edition, Karl S. Lashey, "The Problem of Cerebral Organization in Vision," *Biol. Symposia* 7 (1942), 301. In the second edition, Karl S. Lashey, *Brain Mechanisms and Intelligence* (Chicago: University of Chicago Press, 1929).
27. F. S. C. Northrop, "The Neurological and Behavioristic Psychological Basis of the Ordering of Society by Means of Ideas," *Science* 107, no. 2782 (April 1948): 412. While the first phrase in quotation marks is an accurate quote from Northrop, the second seems to be a free paraphrase.
28. Cf. Wolfgang Köhler, *The Place of Value in a World of Facts* (London: Kegan Paul, 1939), and Raymond Ruyer, *Philosophie de la valeur* (Paris: Armand Colin, 1952).
29. W. Ross Ashby, *Les Mécanismes cérébraux de l'activité intelligente*, in *Perspectives cybernétique en psychophysiology*, trans. J. Cabaret (Paris: Presses Universitaires de France, 1951); and Albert Ducrocq, *Appareils et cerveaux électroniques* (Paris: Hachette, 1952), 144. For an overview, see W. R. Ashby, *Design for a Brain* (London: Chapman and Hall, 1952).

30. Kurt Koffka, *The Growth of the Mind: An Introduction to Child-Psychology* (London: Kegan Paul, 1924), 78 sqq.

31. Kurt Goldstein *Die Aufbau der Organismus* (The Hague: Springer, 1934), passim, particularly page 140. English translation: *The Organism: A Holistic Approach to Biology Derived from Pathological Data in Man* (New York: Zone Books, 1995), *passim*, particularly page 183.

32. TN: "Tendency to orderly behaviour" with "elimination of defects."

33. TN: "The various changes are a unity."

34. Ashby, *Design for a Brain*, chapter III.

35. H. L. Hazen, O. R. Schurig, and M. F. Gardner, *The M.I.T. Network Analyzer* (Cambridge, MA.: MIT, 1931).

36. Ashby, *Design for a Brain*, chapter 8, "The Ultrastable System." TN: This seems to be Ruyer's very free paraphrase rather than a direct quote.

37. Norbert Wiener, *The Human Use of Human Beings* (London: Eyre & Spottiswoode, 1950), chapter 3.

38. Ibid., chapter 3, 69 (first edition); 60–61 (second edition).

39. Ibid., 80 (first edition); 69–70 (second edition).

40. Wiener, *Cybernetics*, 133 (first edition); 113 (second edition).

41. Grey Walter, *The Living Brain*, chapters VI and VII.

42. E. S. Russell, *Le comportement des animaux* (Paris: Payot, 1949), 184. Translation of *The Behaviour of Animals: An Introduction to Its Study*, originally published in 1934; second edition 1938.

43. E. S. Russell, *The Behaviour of Animals: An Introduction to Its Study*, second edition (London: Edward Arnold & Co., 1938), 150.

44. TN: When it was built in 1912, the Nividic lighthouse on the Ouessant island in Brittany was the first automatic lighthouse in the world.

CHAPTER 2

1. Raymond Ruyer, *Elements de psycho-biologie* (Paris: Presses Universitaires de France, 1946), chapter 8.

2. TN: *Unitas multiplex* means "unity in diversity."

3. Helen Marot, *Creative Impulse in Industry* (Boston: E. P. Dutton and Company, 1918), 4.

4. The original of both editions uses square brackets here to indicate the framed part. We use braces instead in order to distinguish from our use of square brackets to indicate original French terms, and bold square brackets to indicate what appears in the first edition only.

5. Robert S. Woodworth, *Psychology: A Study of Mental Life* (London: Methuen & Company, 1922).

6. W. Ross Ashby, "Les mécanismes cérébraux de l'activité intelligente," in *Perspectives cybernétique en psychophysiology*, trans. J. Cabaret (Paris: Presses Universitaires de France, 1951), 8. Translation of Ashby, "The Cerebral Mechanisms

of Intelligent Action," in *Perspectives in Neuro-Psychiatry*, ed. D. Richter (London: H. Lewis and Sons, 1950).

7. Stephen G. Pepper, *A Digest of Purposive Values* (Berkeley: University of California Press, 1947).

8. Pepper, *A Digest*, 7.

9. Raymond Ruyer, *Néo-finalisme* (Paris: Presses Universitaires de France, 1952), 219 / *Neofinalism*, trans. Alyosha Edlebi (Minneapolis: University of Minnesota Press, 2016), 202–3.

CHAPTER 3

1. For example, Wolfgang Köhler writes, "As distance objectively increases, exactly the same thing happens in the brain field physiologically, the increase in distance in the brain field will exactly correspond to the tension which, in a force field, produces a dynamic effect in the same direction." "*As the distance is enlarged objectively, exactly the same occurs in the brain field.**" *Gestalt Psychology* (New York: Liveright, 1929), 390. TN: Another case of Ruyer's free paraphrasing. Köhler does write: "As the objective distance is increased, the corresponding distance in the brain also grows, which is precisely the change implicit in the sense of the vector as given the moment before." *Gestalt Psychology* (New York: Liveright, 1992), 357.

2. Kurt Lewin, *A Dynamic Theory of Personality* (New York and London: McGraw-Hill, 1935) and *Principles of Topological Psychology* (McGraw-Hill, 1936).

3. Lewin, *Principles of Topological Psychology*, 82.

4. Ibid., 33.

5. Ibid., 32–33.

6. TN: The French *actuel* can mean current or present, as well as "actual." This double meaning runs throughout the following sections, and we have translated *actuel* alternatively as "actual" or "present," depending on the meaning which seems dominant in the context. The double meaning, however, should be kept in mind in all cases.

7. TN: Translating the French "*mnémisme*," where *mnéme* means memory in the most general sense. So, something like "memory-ism."

8. The phrase "is believed to be" is in the second edition, where the preceding section on Lewin has been omitted.

9. Lewin, *Principles of Topological Psychology*, 82.

10. TN: Jean-Antoine Watteau, *L'Embarquement pour Cythère* (several versions, 1709–1717).

11. Lewin, *Principles of Topological Psychology*, 16.

12. David Krech and Richard S. Crutchfield, *Théorie et problèmes de psychologie sociale*, 2 vols., trans. H. Lesage (Paris: Presses Universitaires de France, 1952), 88, and *Theory and Problems of Social Psychology* (New York: McGraw-Hill, 1948), 66.

13. TN: Krech and Crutchfield, *Theory and Problems of Social Psychology*, 67.

14. TN: Krech and Crutchfield, *Theory and Problems of Social Psychology*, 69.

15. TN: Krech and Crutchfield, *Theory and Problems of Social Psychology*, 69.

CHAPTER 4

1. First edition: "Pattern" (in English); instead of "Informing Structure."
2. First edition: "Pattern," instead of "Support."
3. TN: "Noegenesis" is a term introduced by Charles Spearman to name the genesis of knowledge according to his theory, which involves inferences made from the comparison of different items of perceptual experience. See Spearman, *The Nature of Intelligence and the Principles of Cognition* (London: Macmillan, 1923).
4. First edition: "Pattern," instead of "Order."
5. Second edition: "impossible to figure entirely in space." Since figure 4.2 is removed from the second edition, references to it are modified.
6. First edition: "Patterns," instead of "Structures."
7. First edition: "Patterns."
8. First edition: "Patterns."
9. Wiener, *The Human Use of Human Beings*, 110; second edition, 97–98.

CHAPTER 5

1. Cf. Edmund C. Berkeley, *Giant Brains* (New York: John Wiley and Sons, 1949), ch. II, 11, sqq.
2. TN. Here is a case where Ruyer is rather free with his quotation, though what he makes Wiener say does seem consistent with his cybernetic principles. Ruyer cites only Norbert Wiener, *Cybernetics, or Control and Communication in the Animal and the Machine* (Cambridge: MIT Press, 1948), 56 (first edition). The passages he has freely quoted from, with a little more context, are as follows: "No operation on a message can gain information on the average. Here we have a precise application of the second law of thermodynamics in communication engineering," Wiener, *Cybernetics*, 91 (second edition). "Thus the modern automaton exists in the same sort of Bergsonian time as the living organism; and hence there is no reason in Bergson's considerations why the essential mode of functioning of the living organism should not be the same as that of the automaton of this type," Wiener, *Cybernetics*, 62–63 (second edition). Ruyer provides this second quote in chapter 7.
3. Harold P. Blum, *Time's Arrow and Evolution* (Princeton: Princeton University Press, 1951), 15 and 23.
4. Cf. Erwin Schrödinger, *What Is Life?* (Cambridge: Cambridge University Press, 1948), and Pierre Auger, *L'homme microscopique* (Paris: Flammarion, 1952).
5. Auger, *L'homme microscopic*, 46. This work is, incidentally, full of very ingenious ideas.
6. TN. To clarify the possible ambiguity of the text here, it is clear from what Ruyer goes on to say that Case III is not the comparison of the first two cases, but simply the case of organized words and phrases.
7. TN. Ruyer's calculation here roughly corresponds with Bell's 7-bit teleprinter code that was the basis for ASCII, the first version of which was developed in 1961. It had 128 code points, 95 of which were printable characters.

8. Raymond Ruyer, *Néo-finalisme* (Paris: Presses Universitaires de France, 1952), chapter XVII / *Neofinalism*, trans. Alyosha Edlebi (Minneapolis: University of Minnesota Press, 2016), chapter 17.

9. Henri Poincaré, *Science et méthode* (Paris: Flammarion, 1920), chapter III: "L'invention mathématique." / *Science and Method*, trans. Francis Maitland (New York: Dover, 2013) (epub), chapter III: "Mathematical Discovery."

10. TN. Poincaré, *Science and Method*, 47.

11. Norbert Wiener, *The Human Use of Human Beings* (London: Eyre & Spottiswoode, 1950), 137–38 (first edition); p. 125 (second edition).

12. As in fact, for example, does Auger in *L'homme microscopique, passim*.

CHAPTER 6

1. Cf. Paul Valéry, *Eupalinos*: "as though acts illuminated by a thought abridged the course of nature; and so we may safely say that an artist is worth a thousand centuries, or a hundred thousand, or even many more!" Paul Valéry, *Dialogues*, trans. William McCausland Stewart (New York: Pantheon Books, 1956), 116. Paul Valéry, *Oeuvres de Paul Valéry* Vol. 1 (Paris: Éditions du Sagittaire, 1931), 140–41.

2. In the second edition, this sentence, the last in the section, is "But one can demonstrate that it is the conscious connections that come first."

3. Clark L. Hull, "Mind, Mechanism, and Adaptive Behavior," *The Psychological Review*, 44, no. 1 (1937): 30.

4. TN. We move here to translating *liaison* as "bond" rather than "connection," since this is the common English term in chemistry, which is the focus of this section. However the continuity with the previous sections should be borne in mind.

5. TN. *Type*—epitome, model, perfect example, archetype. Throughout this section, these various meanings of the French term should be kept in mind.

6. Cf. Raymond Ruyer, "Le 'psychologique' et le 'vital,'" *Bulletin de la Société français de philosophie* 39 (November 1938): 159–95.

7. TN. In the second edition, here and in the following sentence, the "set-up in the direction of the goal" [*Le "montage vers le but"*] replaces "goal-set."

8. TN. Also known as "quantum tunnelling."

9. Louis de Broglie, *Continu et discontinue en Physique Moderne* (Paris: Albin Michel, 1941), 30 and 36; and Pierre Auger, *L'homme microscopique* (Paris: Flammarion, 1952), 97–98.

10. TN. A freak of nature.

11. According to the expression of Gaston Bachelard.

12. Cf. Gaston Bachelard, *L'activité rationaliste de la physique contemporaine* (Paris: Presses Universitaires de France, 1951), chapters II and IX.

13. Cf. H. Dreyfus-Le Foyer, "Les conceptions médicales de Descartes," *Reveue de Métaphysique et de Morale* 44, no. 1 (January 1937): 251.

14. TN. Also called the ductus arteriosus.

15. Cf. Raymond Ruyer, *Élements de psycho-biologie* (Paris: Presses Universitaires de France, 1946), chapter VIII: Les enchaînements substitués.

CHAPTER 7

1. Norbert Wiener, *Cybernetics, or Control and Communication in the Animal and the Machine* (Cambridge: MIT Press, 1948), chapter I.
2. Erwin Schrodinger, *What Is Life?* (Cambridge: Cambridge University Press, 1948), chapters VI and VII.
3. TN. Ruyer uses the term "avant-aprés," which would be more literally translated as "before-after." However, he is commenting throughout this chapter on chapter 1 of Wiener's *Cybernetics*, where the term "past-future" appears (see second edition, pages 34 and 43). We have adopted Wiener's term since it reads at least marginally better in English than the awkward "before-after," and Ruyer likely adopted "avant-aprés" for the similar reason that it reads more naturally in French than *passé-avenir*, although he does also occasionally use this expression, as well as the terms *passé* (past) and *avenir* (future) separately.
4. Wiener, *Cybernetics*, 45 (first edition); pages 34–35 in second edition.
5. Wiener, *Cybernetics*, 53 (first edition); pages 41–42 in second edition.
6. These terms, here and in the next mention in this paragraph, are in English in the first edition only.
7. TN. Wiener, *Cybernetics*, 56 (first edition); pages 62–63 in second edition.
8. Arthur Stanley Eddington, *La nature du monde physique*, trans. G. Cros (Paris: Payot, 1929), 106. Translation of the English original: *The Nature of the Physical World* (London: Macmillan, 1928).
9. Satosi Watanabe, "Le concept de temps en physique moderne et la durée pure de Bergson," *Revue de Métaphysique et de Morale* 56, no. 2 (October 1956): 128–42.
10. Cf. Hans Reichenbach, "Les fondements logiques de la mécanique des quanta," *Annales de l'Institut Henri Poincaré* 13, no. 2 (1952–1953): 109–58.
11. TN. Ruyer's point is lost in translation here, as the English expression "trying not to fall" does not sound as awkward as he suggests the French "*essyer de ne pas tomber*" does.
12. TN. In the original text the first occurrence of "past → present → future" here is *avant → maintenant → après*, while the second is *passé → présent → avenir*.
13. Eddington, *La nature du monde physique*, 53 / *The Nature of the Physical World*, 36.
14. Gaston Berger, "Approche phénoménologique du problème du temps," *Bulletin de la Société française de philosophie* 44, no. 3 (July–September, 1950): 93.
15. TN. The French "actualization" can mean both "realization" or "making actual," and "updating," so linking metaphysical and temporal senses.

CHAPTER 8

1. Émile Meyerson, *L'explication dans les sciences* (Paris: Payot & Co., 1921), chapter X / *Explanation in the Sciences*, trans. Mary-Alice and David Sipfle (Heidelberg: Springer, 1991), chapter 10.

2. TN. Georg Wilhelm Friedrich Hegel, *The Philosophy of History*, trans. J. Sibree (Kitchener: Batoche, 2001), 24.

3. TN. Hegel, *Philosophy of History*, 31.

4. Georg Wilhelm Friedrich Hegel, *The Encyclopdia Logic*, trans. T. F. Geraets, W. A. Suchting, and H. S. Harris (Indianapolis/Cambridge: Hackett, 1991), 238 (section 161).

5. TN. Bergson doesn't seem to use quite this precise expression, but writes, "La vie, c'est-à-dire la conscience lancée à travers la matière" and "La vie, avons-nous dit, transcende la finalité comme les autres catégories. Elle est essentiellement un courant lancé à travers la matière," *L'Evolution creatrice* (Paris: Presses Universitaires de France, 1941), 183 and 265–66. / "Life—that is, consciousness launched through matter" and "Life, I said, transcends finality and all other categories. It is essentially a current launched through matter." *Creative Evolution*, trans. Donald A. Landes (London and New York: Routledge, 2023), 163 and 232.

6. TN. "Nature is not mastered except by being obeyed."

7. Albert Ducrocq, *L'humanité devant la navigation interplanétaire* (Paris: Calmann-Lévy, 1947), 185, sqq.

8. TN. First edition has "divine" [*divine*] instead of "cosmic" [*cosmique*].

CHAPTER 10

1. Cf. L. Landon Goodman, *Man and Automation* (Harmondsworth: Penguin, 1957), 59.

2. As F. H. Allport has emphasized, in critiquing the cybernetic theory of perception (*Theories of Perception*, New York: Wiley, 1955), in the organism, as in the purposive machine, the purpose exists at two levels: (a) my sight guides my hand, searching for a glass of water (secondary purpose, through feedback); (b) but first I am thirsty and I want to drink (primary purpose).

3. See the *Bulletin de l'association de Pédagogie Cybernétique* (Gauthier-Villars), and the special issue of *Europe*, May–June, 1965.

4. Cf. R. Ruyer, *Paradoxes de la conscience et limite de l'automatisme* (Paris: Albin Michel, 1966), p. 34 sqq.

5. Cf. *Cahier du Centre international de Synthèse* III (Zurich, 1956), 10–32.

6. Cf. André Deweze, *Traitement de l'information linguistique* (Paris: Dunod, 1966), 69.

7. TN. See Martin Gardner's "Mathematical Games" column in *Scientific American*, March 1962: "How to Build a Game-Learning Machine and the Teach It to Play and to Win." Reprinted as "A Matchbox Game-Learning Machine," in Martin Gardner, *The Unexpected Hanging and Other Mathematical Diversions* (Chicago and London: University of Chicago Press, 1991).

8. In effect, and contrary to a tenacious assumption, it is not because an electronic calculator makes thousands of multiplications a second that it can perform *any* calculation in a flash. Light may well go very fast (by our human scale), but it still takes more than a thousand years to come to us from the spiral nebula.

9. Deweze, *op. cit.*, 74.
10. Cf. Donald Michie, "Machine Intelligence," in *Penguin Science Survey B*, ed. S. A. Barnett and Anne Mclarren (London: Penguin, 1965), 61.
11. E. M. Braverman, "Essais pour apprendre à une machine à reconnaître des formes visuelles" / "Experiments in Training a Machine to Distinguish Visual Shapes" (in Russian) *Avtomatika I Telemekhanika* 23, no. 3 (March 1962): 349–64. Cited by Deweze, *Traitement de l'information linguistique*, 132.
12. See Richard L. Gregory, *L'oeil et le cerveau: La psychologie de la vision*, trans. Collette Vendrely (Paris: Hachette, 1965) / *Eye and Brain: The Psychology of Seeing*, 3rd ed. (New York: McGraw Hill, 1978).
13. Gregory, *L'oeil et le cerveau*, 172 / *Eye and Brain*, 173: the ellipse seen as a hoop, the shape seen as a puddle, and so on.
14. Cf. Ruyer, *Paradoxes de la conscience et limite de l'automatisme*, chapter XI.
15. Cf. *La Pédagogie cybernétique*, 4 decembre 1964
16. TN. Ruyer's paraphrase of Gilbert Ryle's position in *The Concept of Mind* (Chicago: University of Chicago Press, 1949).
17. Alan Turing, "Computing Machinery and Intelligence," *Mind* 59 (1950): 433–60.
18. Memo 59, R.L.E. and M.I.T. Computational Centre, cited by Donald Michie, "Machine Intelligence."
19. Guido Calogero, "L'homme, la machine et l'esclave," in *Textes des conférences et des entretiens organisés par les Rencontres internationales de Genève 1965 avec le concours de l'UNESCO* (Neuchatel: Les Éditions de la Baconnière, 1965), 43–47.
20. TN. Michael Scriven, "The Compleat Robot: A Prolegomena to Androidology," in *Dimensions of Mind*, ed. Sydney Hook (New York: New York University Press, 1960).
21. Louis Couffignal, *La cybernétique* (Paris: PUF, 1963), 55.
22. François Bonsack, *Information, thermodynamique, vie, et pensée* (Paris: Gauthier-Villars, 1961), 87.
23. TN. The smaller font of this paragraph follows the original. This is a technique Ruyer often used in other works to indicate an example or more marginal point, but it is used only in this chapter of this book, which was added to the second edition.
24. Here and throughout this chapter, as in earlier parts of the book, the word "pattern" is in English in the original.
25. Cf. Ruyer, *Paradoxes de la conscience et limites de l'automatisme*, chapter VI.
26. TN. A reference to Hoyle's Steady State Universe theory, which was a contender to the theory of the Big Bang until the mid-1960s, when the discovery of cosmic background radiation seemed to decide in favor of the Big Bang. According to Hoyle's theory, the universe is expanding and matter is spontaneously created in the gaps between galaxies.
27. Léon Brillouin, *Vie, matière et observation* (Paris: Albin Michel, 1959), 117 sqq.
28. Bonsack, *Information, thermodynamique, vie et pensée*, 103.
29. Gregory, *L'oeil et le cerveau*, 7 / *Eye and Brain*, 9.
30. Gregory, *L'oeil et le cerveau*, 7 *Eye and Brain*, 9.

31. Raymond Ruyer, *La Conscience et la Corps* (Paris: Presses Universitaires de France, 1936); *Néo-finalisme* (Paris: Presses Universitaires de France, 1952) / *Neofinalism*, trans. Alyosha Edlebi (Minneapolis: University of Minnesota Press, 2016).

32. Gregory's translator, Collette Vendrely, naturally suggests, according to current fashion, that this code is of the same type as the RNA code of genetics.

33. Gregory, *L'oeil et le cerveau*, 69 / *Eye and Brain*, 47–48.

34. William Grey Walter, "Physics of the Brain," in *Penguin Science Survey A*, ed. S. A. Barnett, Arthur Garratt, and Anne Mclarren (London: Penguin, 1965), 89.

35. Karl Steinbuch and Uwe A. W. Piske, "Learning Matrices and Their Applications," *IEEE Transactions on Electronic Computers* EC-12, 6 (December 1963), 846–62. Discussed in Deweze, *Traitement d'information linguistique*, 187 sqq.

36. Deweze, *Traitement d'information linguistique*, 197.

37. Cf. Michie, "Machine Intelligence," 75.

38. TN. Restoration to original condition.

39. TN. In English, "the horses gallop." The example here only works in French. The grammatically correct French would be "les chevaux galopent," where the plural is also indicated by "galopent," rather than "galope."

40. Bonsack, *Information, thermodynamique, vie et pensée*, 111.

41. Deweze, *Traitement de l'information linguistique*, 63.

42. TN. Ruyer means that the reader clearly understands that what Bonsack means is "A system left to itself can only lose value," whereas what the printing mistake makes him say is "A system left to itself cannot lose value."

43. TN. "left to itself."

44. Bonsack, *Information, thermodynamique, vie et pensée*, 57.

45. Brillouin, *Vie, matière et observation*, 154.

46. Olivier Costa de Beauregard, *Le second principe de la science du temps* (Paris: Editions du Seuil, 1963), 129.

47. Bonsack, *Information, thermodynamique, vie et pensée*, 146.

48. Raymond Ruyer, *La genèse des formes vivante* (Paris: Flammarion, 1956) / *The Genesis of Living Forms*, trans. Jon Roffe and Nicholas de Weydenthal (London: Rowman & Littlefield International, 2019).

49. Employed by Gonseth for other ends, by the way, in his valuable book *Le problème du temp* (Neuchâtel: Editions du Griffon, 1964), 38.

50. TN. This term can be variously translated as "draft," "sketch," "rough beggings," "preliminary outline," and so on. In the context of biology, its usual English equivalent is "primordium," which refers to a biological structure in its earliest stage of development. In the following, Ruyer uses the term to replace what C. H. Waddington refers to as the egg or the embryo in his descriptions of the "epigenetic landscape." Since Ruyer begins this section with a more general discussion of the term, following Ferdinand Gonseth, we leave this term untranslated at first, then translate freely depending on context.

51. C. H. Waddington, *The Strategy of the Genes* (London: Allen & Unwin, 1957).

52. Cf. Robert Mouton, "Biologie moléculaire et information," *Cahiers internationaux de symbolisme* 7 (1965): 39.

53. Cf. Albert Ducrocq, *Le roman de la vie* (Paris: Julliard, 1966), 203.

54. TN. See S. R. Pelc and M. G. E. Welton, "Stereochemical Relationship between Coding Triplets and Amino Acids," *Nature* 209 (1966): 868–70.

55. C. H. Waddington, *New Patterns in Genetics and Development* (New York: Columbia University Press, 1962).

56. Walter M. Elasser, *The Physical Foundation of Biology* (New York: Pergamon, 1958).

57. Leon Brilouin, *Vie, matière et observation* (Paris: Albin Michel, 1959), m. 25.

58. As Bonsack has demonstrated very well. *Information, thermodynamique, vie et pensée*, 96.

59. For example, G. Sommerhoff, *Analytical Biology* (Oxford: Oxford University Press, 1950). Summarized by Waddington, *The Strategy of the Genes*, 141 and Ducrocq, *Le roman de la vie*, 208 sqq. He affirms that there is a "coupling," but without well explaining how. Waddington himself tries to imagine the possible schema of an *adaptive* genetic mutation, induced by the environment (*The Strategy of the Genes*, 181).

60. Pierre Bertaux, "Les machines à traduire," *Les études philosophiques* 16, no. 2 (April-June 1961): 215–24.

61. L. Couffignal had fun by printing this poem beside a work by Eluard and making the reader guess which of the two poems was the work of a machine. About 30 percent of those tested got it wrong.

62. Deweze, *Traitement d'information linguistique*, 179.

63. TN. Short story by W. W. Jacobs, first published in 1902.

64. Aurel David, *La cybernétique et l'humaine* (Paris: Gallimard, 1965), 132.

65. A remark of Waddington, *The Strategy of the Genes*, 4.

66. Ruyer is referencing a robotic spacecraft, seven of which NASA sent to the moon between 1966 and 1968 in what was known as the Surveyor program, which aimed to study landing on the lunar surface.

67. TN. This by-line appears on the cover of Albert Ducrocq's book *Cybernétique et univers*, vol. 1: *Le Roman de la matière* (Paris: Julliard, 1964).

Bibliography

Allport, F. H. *Theories of Perception*. New York: Wiley, 1955.
Ashby, W. Ross. "The Cerebral Mechanisms of Intelligent Action." In *Perspectives in Neuro-Psychiatry*. Ed. D. Richter. London: H. Lewis and Sons, 1950.
———. "Les Mécanismes cérébraux de l'activité intelligente." In *Perspectives cybernétique en psychophysiologie*. Trans. J. Cabaret. Paris: Presses Universitaires de France, 1951.
———. *Design for a Brain*. London: Chapman and Hall, 1952.
Auger, Pierre. *L'homme microscopique*. Paris: Flammarion, 1952.
Bachelard, Gaston. *L'activité rationaliste de la physique contemporaine*. Paris: Presses Universitaires de France, 1951.
Berger, Gaston, "Approche phénoménologique du problème du temps." *Bulletin de la Société française de philosophie* 44, no. 3 (July–September, 1950): 89–132.
Berkeley, Edmund C. *Giant Brains; or, Machines That Think*. New York: John Wiley and Sons, 1949.
Bertaux, Pierre. "Les machines à traduire." *Les études philosophiques* 16, no. 2 (April–June 1961): 215–24.
Bergson, Henri. *L'Evolution creatrice*. Paris: Presses Universitaires de France, 1941.
———. *The Two Sources of Morality and Religion*. Trans. R. Ashley Audra and Cloudesley Brereton. Notre Dame, IN: University of Notre Dame Press, 1963.
———. *Les Deux Sources de la morale et de la religion*. Paris: PUF, 1984 [1932].
———. *Creative Evolution*. Trans. Donald A. Landes. London and New York: Routledge, 2023.
Blum, Harold P. *Time's Arrow and Evolution*. Princeton, NJ: Princeton University Press, 1951.
Bonsack, François. *Information, thermodynamique, vie, et pensée*. Paris: Gauthier-Villars, 1961.
Boole, George. *An Investigation of the Laws of Thought*. New York: Dover, 1958.
Brain, W. Russell. "The Concept of the Schema in Neurology and Psychiatry." In *Perspectives in Neuropsychiatry*. Ed. Derek Richter. London: H. K. Lewis & Co, 1950.

———. "La notion de scheme en neurologie et en psychiatrie." In *Perspectives cybernétique en psychophysiologie*. Trans. J. Cabaret. Paris: Presses Universitaires de France, 1951.

Brillouin, Léon. *Vie, matière et observation*. Paris: Albin Michel, 1959.

Butler, Samuel. *Erewhon: or, Over the Range*. London: Penguin Classics, 1985.

Calogero, Guido. "L'homme, la machine et l'esclave." In *Textes des conférences et des entretiens organisés par les Rencontres internationales de Genève 1965 avec le concours de l'UNESCO*. Neuchatel: Les Éditions de la Baconnière, 1965.

Cannon, Walter B. *The Wisdom of the Body*. New York: W. W. Norton & Co., 1939.

———. *La sagesse du corps*. Paris: Édition de la nouvelle revue critique, 1946.

Costa de Beauregard, Olivier. *Le second principe de la science du temps*. Paris: Editions du Seuil, 1963.

Couffignal, Louis. *Les machines à penser*. Paris: Les Éditions de Minuit, 1952.

———. *La cybernétique*. Paris: PUF, 1963.

David, Aurel. *La cybernétique et l'humaine*. Paris: Gallimard, 1965.

David, Pierre. *Le Radar*. Paris: Presses Universitaires de France, 1949.

De Broglie, Louis. *Continu et discontinue en Physique Moderne*. Paris: Albin Michel, 1941.

Deweze, André. *Traitement de l'information linguistique*. Paris: Dunod, 1966.

Dreyfus-Le Foyer, H. "Les conceptions médicales de Descartes," *Reveue de Métaphysique et de Morale* 44, no. 1 (January 1937): 237–86.

Dubreuil, Henri. *Standards: Le travail américain vu par un ouvrier français*. Paris: Grasset, 1929.

Ducrocq, Albert. *L'humanité devant la navigation interplanétaire*. Paris: Calmann-Lévy, 1947.

———. *Appareils et cerveaux électroniques*. Paris: Hachette, 1952.

———. *Cybernétique et univers vol. 1: Le Roman de la matière*. Paris: Julliard, 1964.

———. *Cybernétique et univers vol. 2: Le Roman de la vie*. Paris: Julliard, 1966.

Eddington, Arthur Stanley. *The Nature of the Physical World*. London: Macmillan, 1928.

———. *La nature du monde physique*, trans. G. Cros. Paris: Payot, 1929.

Elasser, Walter M. *The Physical Foundation of Biology*. New York: Pergamon, 1958.

Frankfort, Henri. *Before Philosophy: The Intellectual Adventure of Ancient Man*. New York: Penguin, 1946.

Friedmann, Georges. *Problèmes humains du machinisme industriel*. Paris: Gallimard, 1946.

Gardner, Martin. *The Unexpected Hanging and Other Mathematical Diversions*. Chicago and London: University of Chicago Press, 1991.

Goldstein, Kurt. *Die Aufbau der Organismus*. The Hague: Springer, 1934.

———. *The Organism: A Holistic Approach to Biology Derived from Pathological Data in Man*. New York: Zone Books, 1995.

Gonseth, Ferdinand. *Le problème du temp*. Neuchâtel: Editions du Griffon, 1964.

Goodman, L. Landon. *Man and Automation*. Harmondsworth: Penguin, 1957.

Gregory, Richard L. *L'oeil et le cerveau: la psychologie de la vision*. Trans. Collette Vendrely. Paris: Hachette, 1965.

———. *Eye and Brain: The Psychology of Seeing*. 3rd ed. New York: McGraw Hill, 1978.
Grey Walter, William. *The Living Brain*. London: Gerald Duckworth & Co., 1953.
———. "Physics of the Brain." In *Penguin Science Survey A*. Ed. S.A. Barnett, Arthur Garratt, and Anne Mclarren. London: Penguin, 1965.
Hazen, H. L., O. R. Schurig, and M. F. Gardner. *The M.I.T. Network Analyzer: Design and Application to Power System Problems*. Cambridge, MA.: MIT, 1931.
Hegel, Georg Friedrich Wilhelm. *The Encyclopdia Logic*. Trans. T. F. Geraets, W. A. Suchting, and H. S. Harris. Indianapolis/Cambridge: Hackett, 1991.
———. *The Philosophy of History*. Trans. J. Sibree. Kitchener: Batoche, 2001.
Hull, Clark L. "An Automatic Correlation Calculating Machine." *Journal of the American Statistical Association* 20 (1925): 522–31.
———. "Mind, Mechanism, and Adaptive Behavior." *The Psychological Review* 44, no. 1 (1937): 1–32.
———. *Principles of Behavior*. New York: Appelton-Century, 1943.
Jordan, Herman J. "La conception naturaliste du monde dans ses rapports avec la méthode dialectique ou synthétique en biologie." *Recherches philosophiques* 1 (1932): 179–205.
Koffka, Kurt. *The Growth of the Mind: An Introduction to Child-Psychology*. London: Kegan Paul, 1924.
Köhler, Wolfgang. *Gestalt Psychology*. New York: Liveright, 1929.
———. *The Place of Value in a World of Facts*. London: Kegan Paul, 1939.
———. *Gestalt Psychology*. New York: Liveright, 1992.
Krech, David, and Richard S. Crutchfield. *Theory and Problems of Social Psychology*. New York: McGraw-Hill, 1948.
———. *Théorie et problèmes de psychologie sociale*, 2 vols. Trans. H. Lesage. Paris: Presses Universitaires de France, 1952.
Lashley, Karl S. *Brain Mechanisms and Intelligence: A Quantitative Study of Injuries to the Brain*. Chicago: University of Chicago Press, 1929.
———. "The Problem of Cerebral Organization in Vision." *Biol. Symposia* 7 (1942): 301–22.
Latil, Pierre de. *La Pensée artificielle: Introduction à la cybernétique*. Paris: Gallimard, 1953.
———. *Thinking by Machine: A Study of Cybernetics*. Boston: Houghton Mifflin, 1957.
Lewin, Kurt. *A Dynamic Theory of Personality*. New York and London: McGraw-Hill, 1935.
———. *Principles of Topological Psychology*. New York and London: McGraw-Hill, 1936.
Marot, Helen. *Creative Impulse in Industry*. Boston: E. P. Dutton and Company, 1918.
McCulloch, Warren S. *Embodiments of Mind*. Cambridge, MA: MIT, 1965.
McCulloch, Warren S., and Walter H. Pitts, "A Logical Calculus of the Ideas Immanent in Nervous Activity." *Bulletin of Mathematical Biophysics* 5 (1943): 115–33.
Meyerson, Émile. *L'explication dans les sciences*. Paris: Payot & Co., 1921.
———. *Explanation in the Sciences*. Trans. Mary-Alice and David Sipfle. Heidelberg: Springer, 1991.

Michie, Donald. "Machine Intelligence." In *Penguin Science Survey B*. Ed. S. A. Barnett and Anne Mclarren. London: Penguin, 1965.

Mouton, Robert. "Biologie moléculaire et information." *Cahiers internationaux de symbolisme* 7 (1965).

Northrop, F. S. C. "The Neurological and Behavioristic Psychological Basis of the Ordering of Society by Means of Ideas." *Science* 107, no. 2782 (April 1948): 411–17.

Pelc, S. R., and M. G. E. Welton. "Stereochemical Relationship Between Coding Triplets and Amino Acids." *Nature* 209 (1966): 868–70.

Pepper, Stephen G. *A Digest of Purposive Values*. Berkeley: University of California Press, 1947.

Poincaré, Henri. *Science et méthode*. Paris: Flammarion, 1920.

———. *Science and Method*. Trans. Francis Maitland. New York: Dover, 2013.

Reichenbach, Hans. "Les fondements logiques de la mécanique des quanta." *Annales de l'Institut Henri Poincaré* 13, no. 2 (1952–1953): 109–58.

Russell, E. S. *The Behaviour of Animals: An Introduction to Its Study*, 2nd ed. London: Edward Arnold & Co., 1938.

———. *Le comportement des animaux*. Paris: Payot, 1949.

Ruyer, Raymond. *La Conscience et la Corps*. Paris: Presses Universitaires de France, 1936.

———. "Le 'psychologique' et le 'vital.'" *Bulletin de la Société français de philosophie* 39 (November 1938): 159–95.

———. *Éléments de psycho-biologie*. Paris: Presses Universitaires de France, 1946.

———. *Néo-finalisme*. Paris: Presses Universitaires de France, 1952.

———. *Philosophie de la valeur*. Paris: Armand Colin, 1952.

———. *Paradoxes de la conscience et limite de l'automatisme*. Paris: Albin Michel, 1966.

———. *Neofinalism*. Trans. Alyosha Edlebi. Minneapolis: University of Minnesota Press, 2016.

Ryle, Gilbert. *The Concept of Mind*. Chicago: University of Chicago Press, 1949.

Schrodinger, Erwin. *What Is Life?* Cambridge: Cambridge University Press, 1948.

Scriven, Michael. "The Compleat Robot: A Prolegomena to Androidology." In *Dimensions of Mind*. Ed. Sydney Hook. New York: New York University Press, 1960.

Shannon, Claude E. "A Symbolic Analysis of Relay and Switching Circuits." *Transactions of the American Institute Electrical Engineers* 57, no. 12 (December 1938): 713–23.

Sommerhoff, G. *Analytical Biology*. Oxford: Oxford University Press, 1950.

Spearman, Charles. *The Nature of Intelligence and the Principles of Cognition*. London: MacMillan, 1923.

Stapledon, Olaf. *Last And First Men*. 5th ed. London: Metheun, 1934.

Steinbuch, Karl, and Uwe A. W. Piske. "Learning Matrices and their Applications." *IEEE Transactions on Electronic Computers* EC-12, 6 (December 1963): 846–62.

Turing, Alan. "Computing Machinery and Intelligence." *Mind* 59 (1950): 433–60.

Valéry, Paul. *Oeuvres de Paul Valéry*. Vol. 1. Paris: Éditions du Sagittaire, 1931.

———. *Dialogues*. Trans. William McCausland Stewart. New York: Pantheon Books, 1956.
Waddington, C. H. *The Strategy of the Genes*. London: Allen & Unwin, 1957.
———. *New Patterns in Genetics and Development*. New York: Columbia University Press, 1962.
Watanabe, Satosi. "Le concept de temps en physique moderne et la durée pure de Bergson." *Revue de Métaphysique et de Morale* 56, no.2 (October 1956): 128–42.
Wells, H. G. *Men Like Gods*. New York: Macmillan, 1923.
Wiener, Norbert. *Cybernetics, or Control and Communication in the Animal and the Machine*. Cambridge, MA: MIT Press, 1948.
———. *The Human Use of Human Beings*. Boston: Houghton Mifflin, 1950.
———. *The Human Use of Human Beings*. 2nd ed. Boston: Houghton Mifflin, 1954.
———. *Cybernetics, or Control and Communication in the Animal and the Machine*. 2nd ed. (1964). Cambridge: MIT Press, 1991.
Woodworth, Robert S. *Psychology: A Study of Mental Life*. London: Methuen & Company, 1922.

Index

absolute survey. *See* survey, absolute
action, x; compensatory, 38–40
actual, as concept, 65–66
actualization, 71, 125
adrenaline, 54
Aeneid, 96
Alexander, Samuel, 131
amboception, 31
anti-chance, 94–100
Antwerp, 27
Archytas, 16
Aristotle, 17, 65; and the origin of information, 130
artifact, and organism, 132
artificial intelligence, xix
Ashby, W. Ross: homeostat, 38, 39, 41, 55, 71, 144, 152
assembly, xxvi, 171; active versus passive, 47
Astbury, William, 102
astronautics, 134
atom, structure of, 105
atomic bomb, 91
Auger, Pierre, 85
Austerlitz, 152
autocatalysis, 83–84
automata, 8, 14, 28, 45; and automation, 9; in the eighteenth century, 1; and feedback, 137

aviator, 51
Avogadro's number, 93
axiology, 55, 61–72, 137, 141–142; axiological blindness, 58; axiological frame, 56–59; axiological relief, 67–69; and depth, 68; feedback, 141; versus topology, 69

Bacon, Francis, 10
Balzac, Honoré de: *Physiology of Marriage*, 85–86
barriers, ideal, 72; physical versus mathematical, 72
Barthes, Roland, viii
Baudrillard, Jean, viii
behavior, 61–72, 106; and behaviorism, 2
belief, of automatons, 153
Berger, Gaston, 126; and the Société française, 127
Bergson, Henri, 113, 118, 124, 127, 131; supplement of the soul, 8; and time, 118
Berkeley, Edmund C., 8; "Simon" calculating machine, 23
Bernoulli's law, 96
billiard balls, 95
BINAC, 20
binary logic, 81; binary systems, 20

birdsong, 118
Bleriot, Louis, 134
Block, Ned: China brain, xx
Blum, Harold F., 85
Blum, Lesser, 104
body, human, 1
Bohr diagram, 104
Boltzmann, Ludwig, 120, 159
bonds [*liaisons*], 102; chemical, 103; electrochemical, 102; ionic, 105. *See also* connection
Bonsack, François, 155, 159, 161, 167, 169
Boole, George: algebra, 22
Botallo, Leonardo, 111
Boutroux, Émile, 131
Brahe, Tycho, 25, 27
Braille, 4
brain, as a machine, 62
Brain, W. Russell, 35
Brillouin, Leon, xxi, 161, 169, 174
Brownian motion, 94
Buber, Martin, 75
Burckhardt, William: calculation machine, 22, 47
Butler, Samuel: *Erewhon*, 2, 70

Calkins, Mary W., 55
Calogero, Guido, 154
Cannon, D. C. A., 33, 43, 171
Cannon, Walter B., 1
Carnot, Sadi, ii, 6–7, 14, 15, 159
Carroll, Lewis, 178
causation, 125
cells, flip-flop, 12, 20
cephalization, of human industry, 72
chance, 93–112
Chapouthier, Georges, xxi, xxii, 186n51
ChatGPT, xviii
cheating, and learning, 44
circuits, in electronic calculators, 20
cliches, 88
clocks, 1, 126, 127; clockwork, 114, 118
codes, linguistic, 78; theory of, 164
color, 162

Columbus, 72
combinatory, of reason, 77
communication, i–x; 73–80; as the transmission of information, 73
computers, not cybernetic machines, 143
concreteness, principle of, 65–66, 69
conditioning, 40–46
configuration spaces, 61
connection [*liaison*], xvi, 47–48, 93, 97–99, 109; and consciousness, 100–101, 141; improvised, 49; increases information, 98; mechanical versus conscious, 100; as positive anti-chance, 93; between space-time and the trans-spatial, 72; substituted, 49; types of, 113. *See also* bonds
consciousness, *passim*, 2–3, 48, 50, 128; double role, 135; as field, 162–163; as information support, 162–165; and machines, 49–50; nebulous, 48; not needed for communication, 75
contemporaneity, as a principle, 65
continuity, 125; three types, 157
copies, problem of, 160–162
Cornwall, 137
correction, 166–168
cortex, 41, 52
Costa de Beauregard, Olivier, 155, 169
Coué, Émile, 46
Couffignal, Louis P., 1, 151, 156, 159, 169, 178; on "mentality," 155
coupling, 88
Cournot, Antoine Augustin, 135, 142
creativity, 140
crossword puzzle, 90
Crutchfield, Richard S., 71
crystal, 169
cybernetics, *passim*, 1; authentic and inauthentic, 170; as auxiliary of life and conscious intention, 138; and biological morphogenesis, 170–171; confuses memory and trans-spatiality, 77–78; definition, 1, 3–5; fails to understand

information, 141; and informatics, 143–144; information and form, 169; mechanist versus critical, 179–180; myths of, 16–18; postulates of, 5; practical interest, 8–10; reception in France, viii; summary of Ruyer's critique of, ix–xiii; theoretical interest, 10–11

David, Aurel, 178–179
de Vaucanson, Jacques, 16
Deep Blue, xviii
deep learning, xvii
Deleuze, Gilles, viii
delocalization, 175
Demiurge, 116
Democritus, 89, 130, 131
Descartes, René, 10, 19, 109, 130, 181; and the origin of information, 130
detemporalization, 175
determinism, 13
detour, 106–112; microphysical versus conscious, 108
deus ex machina, 37
Deweze, André, 149–150, 168, 179
dialectics, 136; and information, 130
differentiation, 89, 106
DNA, 174
dualism, 138
Dubreuil, Henri, 11
Ducrocq, Albert, 44, 137, 177, 181
ébauche, xxvii, 171–172

Eddington, Arthur, 82, 113, 119, 127
EDVAC, 20
Einstein, Albert, 119
élan vital, 116
Elasser, Walter M., 174
elevator, 55
embryo, xv, 69, 110, 121, 168; and equipotentiality, xv; and induction, 78
embryology, 101, 152, 160, 168, 171–172; Cartesian, 109–110
emotions, theory of, 65

energy, higher forms of, 83
ENIAC, 16, 20
entropy, 82, 119, 122; increase of, 124; negative, 85
epigenesis, 67, 102, 174
equipotentiality, xv
error, 167
esse est percipi, 121
essence, xiv
evolution, 122
expectation, 126
explanation, in Galileo and Newton, 65, 67
eye, 49

fantasy, of inverted times (Weiner), 114–117
Farman, Henry, 26
feedback, 9–10, 29, 20, 40, 43, 75, 146; mechanical, 49–50; molecular, 173; nervous, 67; organic, 31–33; as regulation, 61–62
films, backward, 94–100, 124
finalism, naïve, 66
finality, xiv, 170, 179
Fischer, R. A., 81
flip-flop assembly, 20
Floridi, Luciano, vii, xix, xx
fluctuation, 83–84, 88, 91
Focillon, Henri, 91
force, centrifugal, 29
form, absolute, xiv, xxi; perception of, 148–149
frame, axiological, 56–59
Frankenstein, 16
Frankfort, Henri, 17
Freud, Sigmund, 124
Friedman, Georges, 8

Galileo Galilei, 65, 181
galvanometer, 38
Gamow, George, 157; Gamow effect, 107
Gardner, Martin: *Hexapawn*, 148
genes, as modulators, xv

genetics, molecular, 173
genidentity, 175
geometry, and the divine, 17; non-Euclidean, 70, 89
Gestalt theory, 26, 33, 61–62, 99, 106; Gestalt space, 61
gesture, 62
Gibbs, J. W., 113; statistical mechanics of, 120–121
Gnosticism, and matter, 139
God, images of, 57
Goldstein, Kurt, 39
Gonseth, Ferdinand, 171
Greece, ancient, 17
Gregory, Richard, 162–163
Guattari, Félix, viii
gyroscope, Sperry gyro, 33

habit, 124
Haeckel, Ernst, 66
hands, 10, 49
Hansen, Mark B. N., xix
hartley, as a unit of information, 81
Hartmann, Nicolai, 75
Head, Henry, 36
heads or tails, 96
Hegel, 136; *Philosophy of History*, 130
Heidegger, Martin, viii, x
Henderson, Douglas, 104
Hoffmann, E. T. A., 16
homeostasis, 172; versus homeorhesis, 172
homeotherms, 32
homunculus, 10
hourglass, 126
Hoyle, Fred, 159
Hubel, David H., 163
Hull, Clark L., 23, 33, 101; principle of the "possible robot," 33
Hume, David, 96, 125
Husserl, Edmund, 75; protention and retention, 126

idea, xiv
ideal barriers, 72
idealism, and information, 130
ideals, undetermined, 69–71
images, and after-images, 114
imitation, xviii
in vitro fertilization, 17
inducers, 174
induction, biological, 78
information, *passim*, 83, 143–144; definition, 2; dilemmas of origin, 129–131; information machines, main types, 19–46; meaning of, 77; and novelty, 132; occasional versus structuring, 146; origin of, xxi, 5, 81–92, 129–40; reasoning, 21–23
input, and output mechanisms, 116–117
intentionality, 52–53, 55–57, 135
interests, balance of, 38–40
intuition, 124
invention, 166–168
isomorphism, 89

James, William, 131
Jaquet-Droz, Pierre, 16
Jolly, Friedrich, 110
Jones, Stanley, 171
Jordan, Herman J., on amboception, 31
Jordan, Pascual, 174

Kalin, Theodore, 22, 47
Kasparo, Garry, xviii
Kepler, Johannes, 25
keyboard, 156
Koffka, Kurt, 39, 61
Kohler, Wolfgang, 166
Koran, 146
Krech, David, 71
Kun, Wu, vii

Lamark, Jean-Baptiste, 66
Langmuir, Irving: Langmuir diagrams, 104
language, xvi–xvii, 75–76, 162, 167; as communication, 77
Lapicque, Louis, 1
Laplace, Pierre-Simon, 174

Lashley, Karl S., 37
law of large numbers, 94
learning, 40–46, 141, 146–148; and cheating, 44; self-learning, 179
Leibniz, 10, 20, 121, 142; and the origin of information, 130
Lévi-Strauss, Claude, viii
Lewin, Kurt, 61, 71; analysis of, 64–67
Lobachevsky, Nikolai, 70
localization, problem of, xx–xxi, 81; delocalization, 104, 175
logic, binary, 81
Lorente de Nó, Rafael, 10, 32
Lull, Raymond, 178
Lyotard, Jean-François, viii
Lysenko, Trofim, 176

Mach, Ernst, 26
machines, 19–46; calculating machines, 19–21; inducing machines, 24–27; information machines, xi, 1, 19–46; motor machines, xi, 1; and organisms, 159–160; and reasoning, 76–77; reflex machines, 3; self-regulating, 27–31; teaching, 145; translation, 150–153, 177
Marcel, Gabriel, 75
Mariña experiment, 39
Mark 1 computer, 20
materialism, 72
matrices, in Steinbuch, 165–166
matter, 133, 157
Maxwell, James Clerk, 29; Maxwell's demon, 175
McCulloch, Warren S., 1, 4–5, 11, 35
Mead, George Herbert, 76
meaning, ix, xiv, 2, 48, 63–66, 74, 158, 163; atmosphere of, 22–23; as hyper-spatial, 48–49; and relations, 27
mechanism, and information, 130
mechanism, clockwork, 6
memory, 74, 79; in electronic machines, 77; mnemic self-consultation, 74; as stockage, 77
memory, mechanical, 45

memory, organic, 111
mentality, 152–155
Merleau-Ponty, Maurice, viii, x
Mesozoic era, 9
meta-physics, 129
Meyerson, Émile, 129–130, 137, 138; Meyersonian objection, 140
microphysics, 106–107, 138–139, 175; and cybernetics, 13
mind, 153
Minkowski, Hermann, 120
mnemonic return, 101–102
modelling, xviii
modulation, 14; and modulators, 89
molecules, 103; molecular structures, 104
Morgan, C. Lloyd, 131
morphogenesis, 170–171, 176; of the individual, 171–175; of the species, 176
Morse code, 6, 150
motion. *See* perpetual motion of the third kind
motor machines, 1
Multistat, 144, 152
Musée des Arts et Métiers, Paris, 159
mysticism, 75–76
mythology, 159

Napoleon, 4
Nédoncelle, Maurice, 75
Needham, Joseph, 85
negentropy, x, xxi, 151, 155–156, 159, 161, 169–170, 175
Nemesis, 16
neo-Darwinism, 176
nervous system, 2, 14
Network analyzer, 40
Newton, Isaac, 65, 119, 181; and mechanics, 114; *Principia*, 86
Nividic lighthouse, 45
noegenesis, in Spearman, 27
noise, background, 2
noncontradiction, 96
Northrop, F. S. C., 37

novelty, 138
nuclei, atomic, 104

order, xiii; definition of, 85–87; homogeneous versus structural, xiii
organism, 12; and artifact, 132; constitution of, 136; and machines, 159–60
organogenesis, 171
organs, 110
output, and input mechanisms, 116–117

Paley, William, 83
Pascal, 20, 177–178
Passalong Test, 166
past-future: and activity, 122–125; and entropy, 118–122
patterns, xvi, 2, 156–157; of information, 7
Pavlov, Ivan, 15, 42, 145
pedagogy, cybernetic, 145–146
Pelagianism, 21
Pelc, S. R., 173
pendulum, 29, 97
Pepper, Stephen G., 56–57
perception, 34–38, 100; of forms, 148–149; as mixture, 77
perpetual motion of the third kind, xii, 6–7, 15, 26, 47, 141, 186n10
Perry, Ralph Barton, 56
phenomenology, 76
Philips Company, Eindhoven laboratories of, 43
photography, 10
photosynthesis, 84
physics, classical, 13
physiology, 175
Piccard, Auguste, 135
Pierce, C. S., 131
pipe, 108–109
Piske, Uwe A. W., 165
Pitts, Walter, 4, 35
places, versus states, 105
Plato, xiv, 129, 133; *Meno*, 132, 140; and the origin of information, 130

pluralism, 138
Poe, Edgar Allen, 16
poetry, 4
Poincaré, Henri, 89
possibility, 70–71
Posteraro, Tano S., viii
potential, xvi; state of potentiality, 129–131
pragmatism, 2
preformation, 173
present, the "specious" present, 118
protein chain, 102
psyche, 152
psychology, 3; in Lewin, 65; mechanist, 74–75; positivist, 74–75
Pythagoras, 27; Pythagorean table, 23, 161

radar, 40
rationality, 124
reading, 148–149
reason, and machines, 76–77
reflex, 45
regulation, 142, 169–170; by value, 59
regulator, ball, 30
relation, meaning of, 27
relativity, special, 120
relief, axiological, 67–69
repression, 124
reproduction, biological, xiv–xvi, 79, 102, 132, 171
reversibility, and irreversibility, 93; mechanical, 73–74
Richter, Curt, 152
rocket, V2, 136
Roffe, Jon, viii
runaway, 29
Russell, Bertrand, 17
Russell, E. S., 45
Ruyer, Raymond: critique of cybernetics, ix–xiii
Ryle, Gilbert, on mind, 153

Sauvan, Jacques, 152–153, 164
Scheler, Max, 75

schema, 77; and image, 134
Schoffer, Nicolas, 177
Schrödinger, Erwin, 85, 113, 174
science, xvii, 55–56
Scriven, Michael, 148, 154
Searle, John R., Chinese room, xx
self-awareness, 28
self-regulation, 28
semantemes, 157
sense, xiv, xxvi; sense organs, 2
sensitivity, "interested," 34–38
servomechanisms, 1
servomotor, 28
sets, 50–55; psychic, 123, 127; psychological set, 55, 57; psycho-physiological, 53–54
Shannon, Claude E., vii, xxi, 22, 44, 48, 81, 155
Shepard, R. N., 149–150
signals, 76, 173
signification, xxvi
Simondon, Gilbert, viii
Skinner, B. F., 42, 45, 145
slide rule, 55, 135
space, axiological, 61–72; of behavior, 61–72; configuration spaces, 61
space-time, 175; meaning detached from, xiii, 157–158, 160
Spearman, Charles, 24–26, 168
Spencer, Herbert, 85
Sperry gyro, 33
Spinoza, and eternity, 123; and modes, 131; and the origin of information, 130
Stapledon, Olaf: *Last and First Men*, 17
states, versus places, 105
steam engine, 1, 169
Steinbuch, Karl, 165
Stendhal, 68–69
stereoscope, 27
Stern, William, 55
Strasbourg, 40
structure, 75, 165
stupidity, 165

surface, absolute, 162; "surface-subject," 162
survey, absolute, xiv, 63–64, 100, 121–122, 135, 140; in consciousness, 58, 99–100, 106–107, 120–121; versus step-by-step distance, 63
Swift, Jonathan: *Gulliver's Travels*, 178
symmetry, in communication machines, 74
systems, coupled, 83–84
Szilard, Leo: Szilard's demon, 175

teapot, 133
technics, xvii, 134
thematism, xvii, 64, 74, 89, 140; thematic melody, 36; thematic premonition, 48
thermodynamics, 123; second law of, 82, 120, 121
thermostat, 29, 34, 51–52, 55
Thorndike, Edward, 42, 43
thrust, *a tergo*, 30
time, 123; arrow of, 114, 122–123; dependent on connections, 113; detemporalization, 175; enveloped and enveloping, 126–128; inverted, 114–117
Tolman, Edward C., 34, 56, 101
topology, 160; versus axiology, 69; in Lewin, 64
tortoise, artificial, 34
totalitarianism, 181
toy, pure, 145
trans-actual, 64, 129
transcendent, 58, 59
transitivity, 69
trans-mechanical, 129
trans-physical, xvi, 77, 134
trans-spatial, xiv, 48, 59, 64, 66, 67, 72, 90, 129, 131; versus the supernatural, 61
trans-temporal, 66
truth, 118; truth value, 21
tube [*tuyau*], xxvii, 101, 109; electronic, 4, 11, 117; as a machine that

conserves order, 98; organic, 110–111, 136; lacking in viruses, 139
Turing, Alan, 154
Tycho Brahe, 25, 27

Ulam, Stanislaw, 144
Ultrafax, 79
ultrasound, 10
Umwelt, 45
undetermined ideals, 69–71
uniselectors, 41
unitas multiplex, 138–139
universals, 35, 36, 45, 47, 141; as regulatory, 48

values, as trans-spatial, 67–68, 123
Vauban locks, 40
vectors, 69; in Lewin, 64; vectorial schema, 69
verticality, xvii, xxi, 69, 77
Villiers de l'Isle Adam, Auguste, 16
Virgil: *Aeneid*, 96
visibilia, 90
vitalism, 181
Voder, 23
Vogt and Goertler hypothesis, 102
Voltaire, 83
von Benin, 11
von Neumann, John, 1

Waddington, C. H., 172
Walter, Grey, 34, 45, 147–48, 152, 156, 164; tortoise automaton, 35–36, 38, 44, 55, 58, 144
Watanabe, Satosi, 119, 121
Watchmaker, and the "Winder Up," 83
Waterloo, 152
Watt, James: Watt regulator, 34, 48, 169
Watteau, Jean-Antoine: *The Embarkation for Cythera*, 70
wave mechanics, 80
Weiner, Nobert, vii, ix, xi, xx, 1, 5–6, 35, 42, 45, 78, 81, 85, 91, 113–116, 147, 155, 171; and the time of information machines, 117–118; Weiner's myth, 79–80
Welcon, M. G. E., 173
Wells, H. G.: *Men Like Gods*, 78–79
Wiesel, Torsten, 163
Wurzburg School, 55

About the Author and Translators

Raymond Ruyer (1902–1987) was professor of philosophy at the Université de Nancy. A highly original and prolific philosopher, he sought to provide a metaphysics adequate to the discoveries of contemporary science. Today his works are being rediscovered by a new generation, both in France and beyond. *Cybernetics and the Origin of Information* is his third book to appear in English translation, after *Neofinalism* (2016) and *The Genesis of Living Forms* (2019).

Amélie Berger-Soraruff is research project manager at the Maison Française d'Oxford. She received a PhD from the philosophy program at the University of Dundee, where she also previously taught. Her book *Technics of Existence: Sartre, Foucault, and Stiegler* is forthcoming.

Andrew Iliadis is assistant professor at Temple University in the Department of Media Studies and Production (within the Klein College of Media and Communication). He is the author of *Semantic Media: Mapping Meaning on the Internet* (2022) and coeditor (with Isabel Pedersen) of *Embodied Computing: Wearables, Implantables, Embeddables, Ingestibles* (2020).

Daniel W. Smith is professor of philosophy at Purdue University. He is the author of *Essays on Deleuze* (2012) and coeditor of the *Cambridge Companion to Deleuze* (2012, with Henry Somers-Hall), *Deleuze and Ethics* (2011, with Nathan Jun), and *Gilles Deleuze: Image and Text* (2009, with Eugene W. Holland and Charles J. Stivale). He is also the translator, from the French, of books by Gilles Deleuze, Pierre Klossowski, Isabelle Stengers, and Michel Serres.

Ashley Woodward is senior lecturer in philosophy at the University of Dundee. He is an editor of *Parrhesia: A Journal of Critical philosophy*, and his publications include *Nihilism in Postmodernity* (2009), *Lyotard and the Inhuman Condition: Reflections on Nihilism, Information, and Art* (2016) and the coedited volume *Gilbert Simondon: Being and Technology* (2012).

www.ingramcontent.com/pod-product-compliance
Lightning Source LLC
Chambersburg PA
CBHW021547020526
44115CB00038B/849